T0103482

GLOBAL WIRELESS SPIDERWEB

GLOBAL WIRELESS SPIDERWEB

The Invisible Threat Posed
by Wireless Radiation

VIKAS NEHRU

PARTRIDGE

ISBN:	Hardcover	978-1-4828-8683-2
	Softcover	978-1-4828-8682-5
	eBook	978-1-4828-8681-8

Print information available on the last page.

To order additional copies of this book, contact
Partridge India
000 800 10062 62
orders.india@partridgepublishing.com

www.partridgepublishing.com/india

Contents

For Asmita, love of my life.

Preface

In my otology practice, when listening to patients who reported symptoms such as ringing in the ears and a sense of warmth developing after phone calls, I was embarrassed because I did not have an answer to their problem. It compelled me to pursue the subject further. I audited my work during the years 2005–2014, and I found that I had seen an enormous number of such complaints and majority of patients were young people working in call centres, customer care centres, high-profile business executives carrying at least two mobile phones, and radio jockeys.

In particular, I recall a 39-year-old acoustic engineer who worked in a television studio eight hours a day, six days a week for eight years before he saw me. He was unable to pinpoint what he suffered from. His symptoms didn't seem to fit into a pattern that I could use to make a probable diagnosis. He was constantly ill at ease. The ground under his feet felt wobbly. His ears buzzed. He could hardly sleep. He had lost friends because he was not himself any more. After exhaustive medical investigations, he was handed a diagnosis of central vertigo. No treatment worked.

Intuitively, I asked him to take a month-long vacation in the countryside. When he returned, his symptoms had disappeared. What was happening here? I had to go a long way back in history to find an answer.

Towards the end of nineteenth century, two major inventions—namely, AC electric power and telephone—brought about a paradigm shift that changed the face of the earth forever. A century later, the earth has become shrouded in a veil of man-made electromagnetic radiation. At night in urban areas, sky glow as a consequence of electrification has obliterated the view of the silvery Milky Way.

Before the 1990s, only a few radio and television transmitters emitted microwave radio frequency and were located safely away from public eye. Now an array of microwave-based devices has invaded our homes, offices, and workplaces. Currently, the mobile phones have outnumbered people on earth. Trillions of audio, video, and text messages ceaselessly juggle through our space 24/7, riding on invisible pulsed microwaves. Mobile phone towers are right in your face everywhere. Mobile phone, however, is only the beginning.

In the very near future, we will have an invisible omnipresence of wireless technology invading our physical world. Today the technologies have superseded what could be imagined a few years ago. Having achieved man-to-man and man-to-machine communication, avenues have opened up for machine-to-machine communication. RFID tags with transceivers of their own (of the size no more than a grain of rice) embedded into virtually every object of use will be able to sense, store, receive, and transmit information to and from Internet cloud. The Internet of things (IoT) is expected to have one trillion devices interconnected wirelessly by 2025. Already 40 million devices are connected through IoT in UK alone. Both public and private places will be cluttered with a meshwork of wireless signals. Necessity used to be the mother of invention. Not any more. Now, it would appear, greed is the father of invention.

Whether it is a revolution or pollution depends on whom you ask. But there certainly appears to be a collusion of corporate forces behind it. This raises concerns not just about wiretapping and the human right to privacy but also physical impact on human, animal, and plant health. The World Health Organization has declared low-intensity magnetic fields from power lines as well as microwave radio frequency emissions from wireless devices, including cellular phones, as possible causes of cancer. There is a need for greater public discourse and understanding of the wireless technology and its potential benefits as well as challenges posed by it to human civilization.

At the time of this writing, the National Toxicology Program, under the auspices of Food and Drug Administration of USA, revealed the preliminary results of its $25 million study indicating a significantly increased risk of brain cancer as a result of mobile phone use. We are at a tipping point. It would be imprudent to transfer the cost and consequences to the future.

The speed with which the spider web is expanding globally defies an easy way to redress it. While not pretending to offer a definite solution to the problems posed by this new reality of our lives, this work is an attempt to make explicit the various aspects of that impact. It has led me through a journey across various disciplines of physical sciences, earth sciences, and life sciences, for some of which I had no prior credentials. I have made an honest attempt to present data accurately, but for a work of this nature, which involves interpretation of a large number of studies and theories, it is possible that some errors may have crept in, for which the onus lies entirely with me.

This book has been written without any sponsorship, and I hold no obligation whatsoever to any individual or entity. It was an insurmountable challenge to entirely convert scientific information into a language that a lay reader would find easy to comprehend. It is for this reason that a detailed glossary of terms and few appendices have been included to assist the reader.

If, at the end of it, the reader feels he or she is a step closer in awareness of this new reality of the twenty-first century, I would have succeeded in my task.

Vikas Nehru
19 August 2016

Acknowledgements

Howsoever politically incorrect, I must admit I am grateful to Asmita, my teenage daughter. We would always argue when I insisted she shouldn't sleep with her iPhone under the pillow. I knew I was right, though I didn't know why. Now I do. I thank her for compelling me to stop arguing and find the right reason. I offer this book to her as my answer.

As always, I am thankful to Sushma, my wife, for her brilliantly intuitive inputs but more for letting me lean on her shoulder for continuous encouragement and support.

I can't thank enough Tasveer Kamil for her combative but constructive criticism. Her natural wisdom would shine through her comments every time I asked her to evaluate my work. Her passion for clarity of thought is remarkable.

I am truly indebted to Dr Chidamber Boughram Srinivas and Dr K. R. Sathish, my colleagues and dear friends. They are remarkably skilled surgeons who restored my vision after I had lost it to a detached retina and helped me get back to this work.

Much appreciation goes out to Anna Lissa Rosagaron for her diligent secretarial assistance in putting together a monumental lot of paperwork in an orderly fashion.

Finally, I must thank Pohar Baruah, my publishing associate, without whose help this book wouldn't have seen light of the day.

Chapter 1

Human Experiment without Consent

> Western civilization is a loaded gun pointed at the head of the planet.
>
> Terrence McKenna

Of all human inventions, the mobile phone undoubtedly is one of the most ingenious. The manner of communication has been turned on its head. Wireless communication has provided a wonderful platform for the way we interact, educate, travel, monitor our health, do business, play music, watch movies, capture photographs, and even raise babies.

A small handheld machine that fits in your palm can perform most of the tasks that a bulky desktop computer would do just a few years ago. A mobile phone is no longer an expensive, luxurious toy for the extravagantly rich. It has become as basic to life as utilities, such as water and electricity. Machine-to-machine (M2M) communication is fast erasing the need for empathy in human interactions.

The rapid growth of smartphone apps will ensure that there will be no end in sight to this revolution.

RFID (radio frequency ID tags) is being rapidly adopted for consumer goods for logistics and marketing. No more do you need to swipe a credit card to make payments. You can simply wave your mobile phone near a cash register, and the transaction will be made by near-field communication (NFC). Ever-increasing download speeds of wireless Internet services offered by 4G and LTE (Long Term Evolution) and Android tablets and Apple iPads mutually complement each other. It is estimated that by 2020 there will be 50 billion devices connected wirelessly and by 2025 the figure could reach 1 trillion. Wireless sensor networks (WSN) will soon automate several activities inside homes.

While we love our devices, we are almost never cognizant of an invisible and intangible infrastructure of wireless radio frequency electromagnetic waves that drives them. If it were possible to look down from space with an electromagnetic sensor, one would see an enormous web of electromagnetic fields enveloping every inch of the planet. It is as if another layer has been added between the planet and the ionosphere.

Telephony has become cheaper over the years because there are so many willing users. If you can invent a product that everybody is queueing up to buy, then bringing down the cost of production is no big deal—not to speak of the sophisticated spin given by deceptive advertising, and the strong arm of nexus between policy makers and industry! While clean drinking water is still scarce in more than 40% of the inhabited world, mobile phones have already exceeded the human population. That says a lot about the priorities of our political and economic systems.

Traditionally, a formal pursuit of science was never intended to develop technology. The motivation of scientific pursuits in its purity is nothing but to understand the mysteries of nature. 'I want to know God's thoughts; the rest are details,' said Albert Einstein. In that endeavor, there is information and knowledge gained and universally shared.

The foundations of the world of telecommunication as we know it today were laid from nineteenth century onwards. Long-range telegraph arrived in 1837 (Morse), the telephone in 1876 (Bell), commercial electric networks in 1880 (Edison), induction motor in 1892 (Tesla), wireless receiver in 1896 (Marconi), and before the mid twentieth century, radio had arrived. High frequency radio frequency was introduced in 1950s as FM radio and television.

Unlike today, there was hardly any military funding into science. Credit for development of science and technology in leaps and bounds in the last century goes almost entirely to governmental military funding. Civil benefits are only by-products of the process. Even Galileo had tried, albeit unsuccessfully, to court the attention of the military-minded Republic of Venice before he turned his telescope to astronomical research. Twentieth century brought in its wake two world wars. Chlorine, a powerful poison gas used in World War I, was a contribution of the German dye industry. In World War II, British and American work on developing microwave-based wireless communication systems led to development of radioactive detection and ranging (radar).

Not only the democracies but also corporations in alliance with governments flourished. Science moved its domain from laboratories to our kitchens and living rooms. So powerful was its influence that science, both political and otherwise, laid a new set of rules that completely effaced the authority of beliefs held until then.

Consequent upon the invention of the electric bulb by Thomas Edison in 1880 and alternating current by Nikola Tesla a few years later, infrastructure for electrification was laid down with great fervour all over what we know as the developed world today. The world was ready to go to war. And it did. Although separated by just two decades in chronological time, the Second World War was fought with enormously improved military and technological arsenal—thanks to science.

Though industrial revolution had already been ushered in long before the wars, the fillip it got after the wars was unprecedented. Beginning in the nineteenth century and coming to fruition in the twentieth century, major changes occurred in all walks of life. High-voltage power lines were already in place. Information and communication technology (ICT) boosted the capabilities of warring nations. Its eventual transition from analogue to digital raised the bar even further.

There has been an incremental addition of man-made electromagnetic radio frequency to the environment through the latter half of twentieth century in the shape of FM radio, television, radar for military as well as civilian use, undersea communication cables, microwave ovens, cellular telephony, Internet, wireless devices, and what have you. The intensity of man-made low frequency electromagnetic exposure alone of the current generation is of the order of one followed by eighteen zeroes.

Perhaps it wasn't conceived at that point of time that extending the wireless communication system to civilian use would bring enormous dividends both in terms of conclusively entrenching a place for modern science in the hearts and minds of the entire human race, who were still in a state of stunned shock after having just seen the destructive face of science in World War II, as well as the enormous potential for commerce that would transform the economies of the world. But that is precisely what followed.

Baby boomers hadn't anticipated this. There are parallels one could draw between the explosion of the human population and technology. In AD 1 population is estimated to have been between 200 to 300 million. It took 18 centuries for it to grow to 1 billion in 1804. The technology entered the human civilization, and the second billion was achieved in 125 years (1930), the third in less than 30 years (1959), the fourth in 15 years (1974), and the fifth billion took just 13 years (1987). We were 6 billion at the dawn of the twenty-first century, and in January 2016, we had crossed the 7.4 billion mark.

According to the live tracker of the GSM Association, the number of wireless devices exceeded 7.8 billion in July 2016 and is still counting. Wireless communication systems have arrived and are here to stay forever. Presently, the success of automated intelligence, surveillance, and reconnaissance assets of nations depend on access to the global wireless broadband industry. The ever-increasing congestion and competitiveness has led to new monsters such as cyberspace crime and consequently necessitated cyberspace policing.

Percivall Pott, an English surgeon, was the first to describe the relationship between human disease and environment. In 1775 he found a link between cancer of the scrotum in chimney sweep boys and soot emanating from chimneys. Industrialization has come a long way since then. Now we have a host of environmental agents linked to serious health problems. All that we need for basic survival comes packaged in the resources of the planet. However, being an advertisement-driven, addictive society, we set no limits to our consumption beyond the means of basic survival. The looming climate crisis due to greenhouse gases, ozone depletion, and acid rain is the culmination of series of events related to industrialization in the past two centuries.

Evidence indicates that the increase in the incidence of cancer has a paralleled increase in industrialization in the past hundred years. While part of it may be due the fact that average lifespan has increased in the last century, it is without a doubt that exposure to new and toxic environmental agents has contributed to the rise. From a biological standpoint, the fundamental change that initiates cancer is damage to

DNA. There is a constant war between DNA damage on one hand and DNA repair on the other. Cancer initiates when enough damage has accumulated beyond the capacity of the body to mount repair. What is new is that we have generated a vast number of agents, leading to exposure from a very early age, which leads to DNA damage.

For instance, sulphur dioxide is one of the major particulate pollutants in the air. Its main source is coal power plants and oil refineries. Inhalation of sulphur dioxide is linked with asthma and heart disease. Dangerous levels of formaldehyde, a known carcinogen, have been seen in wastewater and airborne emissions from fracking operations. Benzene, yet another carcinogen, is a solvent used in energy industries, such as coal, oil, and gas. Coal-based power plants emit mercury, known to be toxic to the nervous system. Add to them toxic chemicals, pesticides, and insecticides, and you have a recipe for disaster.

Many people become sensitive to low levels of noxious emissions from toxic chemicals and develop a condition known as multiple chemical sensitivity (MCS). It is also known as idiopathic environmental illness. As we shall see later, yet another cause of environmental sensitivity is our exposure to electricity and electricity-driven appliances. People who develop multiple chemical sensitivity (MCS) usually also develop electrohypersensitivity (EHS).

While on one hand, science reveals the impact of invisible agents in our environment, which are detrimental to health of all living species, industry employs a two-pronged strategy to achieve its goals in corporatizing goods and services. For one, it produces a scientific counternarrative by funding its own research with the help of for-hire scientists. Secondly, such is the impact of industrial power that lobbyism has become a parallel profession to influence the process of policy making and deflect it from public interest to the interest of corporations. The governments are compelled to walk a thin line between public interest and corporate interest and sadly have often leaned towards the latter. Big Tobacco is a classic example that misled societies for over half a century before restrictions were legalized.

Ubiquitous pervasion of wireless technology defies all ten points of the Nuremberg Code, which was set out in Nuremberg, Germany, in the wake of the Second World War and made voluntary, well-informed consent of human subjects in any scientific experimentation an absolutely essential prerequisite. Advertisements are calibrated to the advantage of corporations. They never tell you the flip side of consumer products. Dazzled by the products, we foolishly make uninformed choices.

The scale of impact on all forms of life is enormous. Living species have never had to learn to adapt to changes in environment with such an abrupt speed. The human race is challenged with the need for making a giant evolutionary leap for biological systems to learn to cope with man-made wireless microwave radiation. But alas, evolution is an extremely slow process.

What is most amazing and perhaps frightening is that neither society nor science nor politics are prepared for the wireless revolution, which has descended upon us as an invisible demon everywhere around, and what it could do to us appears ominous. Being short-sighted doesn't help. Now that the genie is out of the bottle, there is a need for departure from business as usual unless ways are found to create biocompatible signals to power communication technology.

All that the regulatory authorities can depend on is the scientific evidence that exists today. The industry and its scientists manufacture doubt by disputing every scientific study that argues against the safety of mobile phones. If science could move forward without the industry creating hurdles, the proof could come sooner. Waiting for absolute proof may be too late. What you need to look at is not absolute evidence but the direction in which the existing evidence is pointing. It takes years to fill gaps in scientific knowledge. This is particularly so in case of environmental toxins that affect humans because ethical considerations would not allow scientists to conduct experiments on humans.

In the case of the fast-food industry, the economic burden of the cost of treating obesity and diabetes outweighs the revenues raised by marketing fast food, making it thereby a poor business model from a nationalistic point of view. Whether the same or even worse will happen in case of microwave revolution remains to be seen unless, of course, it is by a deliberate design, a theory that has been doing rounds.

The guiding principle of sharing knowledge is the benefit of mankind as a whole. But we have among ourselves a certain species that sees the potential for economic gains from that knowledge and can set all scruples aside so long as it serves the 'purpose'. A cursory glance at the Forbes list of richest people on earth would tell you that of the top ten on that esteemed list, at least 40% have their businesses riding primarily on the electromagnetic waves in the range that is the subject of this book. Lobbyists get laws written in their favour. Therefore, there is little room for surprise why the 'waves' won't go away.

With the advent of the Internet of things, the agenda appears to be to engineer further course of the future by replacing real life by a virtual world. Besides, now that the corporations represent unelected behind-closed-doors members of governments, it will be no surprise when (not *if*) our lives become robotized numbers on a binary scale. Reports such as Silent Weapons for Quiet Wars,* if true, might just stand testimony to this.

Declassified documents in the public domain testify to the fact that unethical experiments had been performed in twentieth century on human subjects, such as exposing people to chemical and biological agents, using mind-altering substances to extract truth from prisoners to prostitutes and captured spies (MK-ULTRA), feeding radioactive material to mentally disabled children, exposing testicles of prisoners to high doses of radiation, and many others. The search was to find a 'truth serum'. It was never found.

On 10 March 1995, President Bill Clinton in his public apology to the nation said, 'Thousands of government sponsored experiments did take place at hospitals, universities and military bases around our nation. The goal was to understand the effects of radiation exposure on the human body. While most of the tests were ethical by any standards, some were unethical, not only by today's standards but by the standards in the time in which they were conducted. They fail both the test of our national values and the test of humanity. Informed consent means the doctor tells you the risk of the treatment you are about to undergo. In too many cases informed consent was withheld. Americans were kept in the

---

* On July 7, 1986, an employee of Boeing Aircraft Co. bought scrap that included a photocopier. Inside, he found a copy of the document entitled 'Silent Weapons for Quiet Wars: An Introduction Programming Manual'. It details plans for gaining control over societies en masse by employing weapons that do exactly what weapons do but silently.

dark about the effects of what was being done to them. The deception extended beyond the test subjects to encompass their families and American people as a whole for these experiments were kept secret. And they were shrouded not for compelling reason of national security but for the simple fear of embarrassment, and that was wrong.'

Is the current use of microwave radiation human experiment version 2?

There are lessons we could learn from the past. It wasn't until the 1920s that the hazards of smoking began to be understood. Up until the 1930s and 1940s, incidence of lung cancer was rising. Persuasion of governments by the scientific community about health hazards of smoking went unheeded for nearly half a century. It wasn't easy to stand up to Big Tobacco. The advertising industry and Big Tobacco were mutual benefactors and beneficiaries. Indeed, the business of advertising learnt its initial lessons from advertising tobacco products. In the early part of the nineteenth century, smoking was a taboo for decent women. Until World War I, if a woman smoked in public, she was taken to be a prostitute. Advertising changed that. What with celebrity women depicting smoking in advertisements, holding up a cigarette became a symbol of confidence and freedom. Cigarette became the hottest product for the advertising industry. Celebrities were seen on advertisement hoardings, espousing cigarettes and even claiming that the smoke cleaned their lungs! Even Santa Claus was roped in the business of advertising.

It took fifty years of persuasion and lawsuits before it became a statutory requirement to warn the end user through a label on every single pack of cigarettes. It could not have been otherwise because nearly one-third of revenues in the Unites states came from tobacco taxes in the early part of the last century.

We don't see blatant advertisements any more. Smoking usually begins in the teenage years or adulthood. Exposure to EMF radiation on the other hand, may begin even before the child is born!

Even though the first formal diagnosis of asbestosis was made as early as 1924, the first lawsuits didn't appear in courts until the 1970s. The governments and the industry knew it, but they concealed the fact from the public. It wasn't until 1989 that asbestos began to be banned. The World Commission on the Ethics of Scientific Knowledge and Technology (COMEST) reported in 2005 that an estimated 250,000 to 400,000 deaths from mesothelioma, lung cancer, and asbestosis due to past exposure to asbestos would occur between 2005 and 2030.

That was a tragedy that didn't need to happen.

Dr. Robert Becker, noted for decades of research on the effects of electromagnetic radiation, has warned, 'Even if we survive the chemical and atomic threats to our existence, there is the strong possibility that increasing electro-pollution could set in motion irreversible changes leading to our extinction before we are even aware of them. All life pulsates in time to the earth and our artificial fields cause abnormal reactions in all organisms . . . these energies are too dangerous to entrust forever to politicians, military leaders and their lapdog researchers.'

When you raise children, you want them to trust you with having created a safe environment for them to flourish. Can we justify their trust?

Chapter 2

Our Planet

> If we try to pick out anything by itself, we find it hitched to everything else in the universe.
>
> John Muir

Life on earth has been shaped over the millennia by forces of Mother Nature. Depending on the distance over which a force is able to exert its influence, the four fundamental forces of nature, in the order of strength, are strong nuclear force, electromagnetic force, weak nuclear force, and finally, gravitational force. Much like the dexterity of a puppeteer, the cosmic dance of these forces simultaneously has dynamic elegance and coherence.

Our universe began 13.73 billion years ago. How do we know this? Well, what we do know from observations in astrophysics and cosmology is that the universe is constantly expanding. Its constituent stars and galaxies are moving apart from one another. That being so, it is safe to assume that, at some time in the past, they must have been close enough to have arisen from a singular source. From the calculations of the speed at which they are moving apart and their relative distances from one another, it is possible to surmise the point in time at which it began from the singular source, and that turns out to be 13.73 billion years ago, when universe (and indeed space and time) began from an explosion in a singularity of the size of a billion-trillion-trillionth of a centimetre called Big Bang. The universe at that time was violently hot. At this point in time, all four fundamental forces of nature were unified. This was followed by the expansion and cooling of the universe.

Expansion led to the separation of forces, resulting in an early soup of subatomic particles coming together to form the building blocks of all matter—namely, electrons, protons, and neutrons. At about 380,000 years after the Big Bang, the universe had cooled enough to allow negatively charged electrons to attach to positively charged protons to form electrically neutral hydrogen atoms and subsequently two hydrogen atoms to coalesce to form helium atoms. This made it possible for electromagnetic radiation to move freely. The universe was no longer a dark dungeon. It was illuminated!

Expansion and cooling continued for millions of years that followed. Thus, from very high frequency radiation was born low frequency radiation. Cosmic microwave background is the remnant of cooled radiation from 380,000 years ago after the Big Bang that still pervades the universe. Eventually, expansion had slowed down, but in the last few decades, according to observations of brief stellar explosions called supernova, the expansion is now again accelerating. It is yet to be understood why it is so. Whether it will continue to expand or collapse again into big crunch is not yet known. Unless this is discovered, the future of the universe cannot be predicted.

Planetary Forces Join to Create Life

Life evolved in the womb of planetary forces, such as electromagnetic visible and infrared solar radiation, extremely low frequency resonances, subatomic nuclear forces, gases such as carbon dioxide and oxygen and ozone, water, gravitation, and geomagnetism.

Solar Radiation

For 4.57 billion years, the sun has been the primary extraterrestrial source of electromagnetic radiation sustaining all life on the planet. Without radiation from the sun, the earth wouldn't exist, and if it did, it would be as cool as interstellar space (about -270 °C).

It takes about eight minutes for electromagnetic radiation emitted from the sun to travel 149.6 million km to reach the Earth. The major fraction of the sun's radiation is in the 'visible' part of the electromagnetic spectrum; however, the sun radiates electromagnetic energy over a very wide range of wavelengths.

Solar radiation, though, is a double-edged sword. On one hand, it is necessary in keeping us warm through its infrared segment, in allowing us to 'see' in the visible-light portion of its spectrum, in enabling photosynthesis, in regulating circadian rhythms, and in helping us synthesize vitamin D. On the other hand, UV radiation emanating from the sun can cause damage to living species. With its energy output of 3.86×10^{26} J/s, our sun sends out X-rays and solar winds towards Earth. Solar winds comprise a stream of highly charged particles hurled towards the Earth. If it were to reach the surface of the Earth, it would seriously damage living species of all kinds. Fortunately, the Earth's magnetosphere deflects most of it.

Cyanobacteria Set the Ball Rolling

Primordial planet Earth was formed 4.54 billion years ago. It is the only planet known to have water on its surface and oxygen in its atmosphere to allow life to flourish.

Our ancestor is not the ape. By the time apes appeared on Earth, the evolution of species had gone a long way up the ladder. Our real ancestors are anaerobic microbes growing in the bottom of the ocean called cyanobacteria or blue-green algae. Over 3 billion years ago, cyanobacteria were the first molecular machinery to use sunlight and carbon dioxide to synthesize its food through a process that was later termed as photosynthesis. The only problem was the by-product of the process was oxygen, which at that time ironically was indeed a toxic waste. The oxygen partly got absorbed into the water in the oceans and partly escaped into the atmosphere. In the ocean, it oxidized ores of iron produced by volcanic eruptions in the ocean floor, and the atmosphere became rich in oxygen, allowing life to flourish and even providing a protective shield from solar radiation by creating a blanket of ozone layer.

We live on a timescale of years and decades. Evolution takes place over millennia, as discovered by piecing together millions of evidences such as fossils, DNA, comparative structure, and function of various species. The blue-green algae changed the atmosphere of our planet by enriching it with oxygen as a result of photosynthesis over 3 billion years ago; the first nucleated cells, 1 billion years ago; the first vertebrates, 500 million years ago. It took nature another 2.25 billion years to get just the right balance between gravitation, strong and weak nuclear forces, and electromagnetism in order to prepare the planet for arrival of the *Homo sapiens*, supposedly its most evolved species. Though our predecessor *Homo* had arrived 2 million years ago, man, in his current anatomical form, arrived about 200,000 years ago. The truth is that all living species are cousins of one another.

Gravitation

Gravitation pulls all matter together. It has remained unchanged and can operate unimpeded by anything. It must be pointed out though that not everything is yet understood about gravity. The quantum theory proposes graviton, akin to electron of EM waves, as basis of gravity. Graviton, though, is yet to be discovered.

When life evolved from the ocean to the earth, the amphibians and reptiles consumed a lot of energy to slither along the surface of the earth as the entire body experienced the pull of the gravity. An erect stance would beat the gravity, but that would require a new kind of adaptation. As species became erect, the need to rearrange dynamics of body fluids against the magnitude and direction of the gravitational pull required evolutionary readjustments in the musculoskeletal organization of the body. Weight of an object equals its mass times the gravity. Biological processes vary according to the weight of the body. We are all aware of the perils of obesity looming as the next big epidemic. If gravity can influence weight and therefore biology, then it must be a major force in our environment, shaping life on earth.

Earth Is Magnetic

Omnipresent geomagnetic field (GMF) is a prerequisite for life. The existence of the geomagnetic field preceded the evolution of living species. Therefore, it is conceivable that the atoms and ions that make up living organisms would interact with magnetic forces in its environment.

Earth's magnetic field, which causes a compass to point north, is a static field originating from the metallic component of Earth's crust and the flow of direct current in the liquid part of Earth's core. Magnetosphere extends from Earth's interior towards space 30,000–50,000 mi. from the surface of the Earth. It exerts a stronger influence than the solar or interplanetary magnetic field and thus prevents the solar wind from stripping away the ozone layer. It has been recorded to be lowest at the equator, at 30 μT (300 mG), and highest at the poles, at 70 μT (700 mG). There is geological evidence that polarity of Earth's geomagnetic field reverses at intervals of thousands of years. At the level of the seabed, the geomagnetic field may be perturbed by sunken ships, mineral deposits, telecommunication cables by up to hundreds of nanotesla. With the exception of geomagnetic storms, the intensity of electromagnetic activity in nature is constant within a certain pattern of periodicity.

Magnetoreception is a sense that allows an organism to detect geomagnetism in order to orient itself in terms of direction, altitude (as in birds), and location. It has been demonstrated in bacteria, fruit flies, lobsters, honeybees, birds, turtles, sharks, and stingrays. Migratory birds detect the earth's magnetic field via iron-rich magnetite-like structures in their upper beak in order to navigate through their long journeys. It tells the migratory birds how to find their destination and tells the roots and branches of a plant in which direction to grow. The average intensity of the static electric field of the earth is 130 V/m, and the magnetic field is 0.5 G. A variation as small as +/– 0.1 G during a geomagnetic storm can cause serious health effects on animals and human beings, such as nervous disorders, high blood pressure, and even heart attacks.

Biofield

Nuclear forces operate inside the nucleus of the atoms, keeping subatomic particles together. In the last few decades, a new force named as spin is said to be present around all bodies. In case of living organisms, this force is particularly large and is called biofield, an abbreviation for *biological energy field*.

When you have an ECG or an EEG performed in a hospital, the test result is shown as a graphic output of bioelectric activity of cells. This is so because biological machine employs electric and magnetic fields in its moment-to-moment functioning at cellular and molecular level. Indeed, some of that energy emanates from the body of living organisms. It is referred to as resting potential and can be quantitatively measured. The ionic concentration inside and outside of a cell varies, creating an electric potential difference in the range of 50–80 mV. During excitation as simple as moving your finger, it can jump up to 120 mV as it moves from resting to action potential. This is the very foundation of life.

The use of the ECG and EEG for over half a century in the field of diagnostics in medicine is an acknowledgement of the existence of the biofield. However, the truth goes far beyond that. The biofield is essentially a sum total of energy emanating from all living species in the form of electromagnetic field, light, heat, and sound at extremely subtle levels.

The human heart constantly pours out a powerful electromagnetic field that can be recorded in the region of space several feet away from the body. Its electrical component is 60 times stronger, and its magnetic component, as measured by superconducting quantum interference device (SQUID) based magnetometre, is 5,000 times greater than that of the brain. Magnetic fields from heart and brain can be measured as magnetocardiogram (MCG) and magnetoencephalogram (MEG) respectively. It travels outwards and can interact with fields emanating from other sources of electromagnetic fields, which include other people, other living species, geocosmic fields, and indeed, the entire biosphere.

Bioelectric fields are generated within the body from every organ, but particularly strong fields are generated by the heart and brain. Some marine organisms, such as freshwater eel, use these fields to hunt prey, avoid predators, and grope for opportunity for mating. From its 6,000 specialized cells called electrocytes, an electric eel can produce a 600 V electric field, which is five times compared to what comes out of a wall socket. Measurements of biofield have been done and have shown that it can vary with changes in the earth's geomagnetic field.

Optical Radiation

Initially, the visible light portion of the spectrum was all that was known for obvious reasons. Science has since shown that light is just the visible part of a huge part electromagnetic spectrum that constantly surrounds us but is invisible to humans. Some birds, it has been found, can see some part of ultraviolet range as well.

Ultraviolet light forms the high-energy component of the solar spectrum. Ultra-violet light has a wavelength *shorter* than those in the visible part of the solar spectrum and is not detected by human vision. Most of the sun's ultraviolet radiation, which has very short wavelength, is filtered out by Earth's ozone layer.

Also referred to as thermal radiation, infrared radiation is invisible to the human eye and has a wavelength *longer* than those in the visible part of the solar spectrum. Water vapour in the earth's atmosphere effectively blocks most of the sun's incoming infrared radiation. At the longest infrared wavelengths, the earth's atmosphere is somewhat more transparent; nevertheless, it is the atmosphere's opacity to infrared radiation that is largely responsible for keeping the earth warm. The natural electromagnetic spectrum also includes gamma radiation of cosmic origin and radioactive elements, such as uranium and radium.

Schumann Resonances

The ionosphere is created as the sun's radiation breaks up gas molecules in the thermosphere to create positively charged ions. It has three layers, the lowest at about 90 km above sea level, known as the D region; next is the E region, which peaks at about 105 km, and then the F region, which peaks at about 600 km and absorbs very high frequency UV radiation. It reflects man-made electromagnetic waves, thus making radio transmission possible. D and E regions reflect AM radio waves, whereas F region reflects shorter wavelengths.

As the ionosphere's electrical currents are influenced both by radiated forces of the sun and the moon, it impacts the electric field of the earth as well. It has been estimated that approximately 1,000–2,000 thunderstorms and about 3,000–5,000 simultaneous bursts of lightning strikes happen globally at any given point of time. As they strike the surface of the earth, a massive amount of electrons are absorbed by the earth's surface, causing it to become negatively charged.

It would appear from the above that the earth and the ionosphere are quite like two concentric balls with one placed within the other; in this case, the earth is negatively charged, and the ionosphere is positively charged. There must exist an electric tension between these two. In physics, two concentric balls with opposite electric charges are known as ball condensers.

In 1952, Professor W. O. Schumann (1888–1974) of Technical University of Munich, Germany, during a lecture on electricity, asked his students to calculate the frequency of electromagnetic waves generated in the cavity between two such balls. They came up with a figure of 10 Hz. That, as was to be found later, was close to the frequency of background electromagnetic waves in our atmosphere!

Later in 1960, these were actually recorded by Balser and Wagner, using a high metallic mast as a vertical electric antenna at the campus of MIT Cambridge, Massachusetts, USA.[†] Hans Berger had already recorded the first EEG in 1924. While reading the results of Professor W. O. Schumann in a journal, Dr Anker Muller, a German physician, and a colleague of Hans Berger, was struck by the amazing similarity between the earth's background electromagnetic frequency and the frequency of alpha waves in EEG of human brain when human mind it is in harmony.

---

† A. P. Nickolaenko and M. Hayakawa, *Resonances in the Earth-Ionosphere Cavity* (2002).

He persuaded Schumann to examine the coincidence. The task was assigned to one of Schumann's students named Herbert Konig (who later succeeded Schumann at the Munich University). He compared background electromagnetic frequencies in our atmosphere and human EEG recordings and found that the fundamental frequency of natural oscillations was 7.83 Hz, which is exactly similar to that of alpha waves of the human brain! Quite aptly, they are now known as Schumann resonances. These are essentially a weak, extremely low frequency electromagnetic field modulated by the micropulsations and sculpted by the solar and lunar cycles.

That we are coupled to a global wavelength was a giant step in our understanding of the relationship between the atmosphere and evolution of life on earth. In his book *Bioinformative Medicine*, Wolfgang Ludwig refers to ancient Chinese texts which indicate that mankind depends for survival on two forces—one from above (*yang*) and another from below (*yin*). He believes that these were actually references to Schumann resonances from above and geomagnetism from below. Other texts have pointed to the similarity between the Hindu mantra 'Aum' or 'Om' as being Schumann resonance or the voice of nature.

Like an enormous universal hum, Schumann resonances occur at 7.8 Hz, and its multiple harmonics occur at 14, 20, 26, 33, 39, and 45 Hz. The spontaneous frequency of a mammalian brain, as shown by EEG, is at 8–14 Hz when the mind is relaxed, and during concentration, it is at 14–30 Hz. When we are in deep sleep or in deep meditative state, our brain is in delta state (1–3 Hz). This is when the body rejuvenates and the cells repair themselves. When we are in theta-alpha state, we are still very calm and relaxed. When fully alert, we are in beta state. At any one point in time, one of the frequencies is more dominant than the others. It appears that our own electromagnetic field is coupled with the electromagnetic field of the universe at 7.83 Hz. At this coupling, the brain waves are in alpha-theta range, which is a natural state of meditative joy. If you encounter very high frequency electromagnetic waves, your brain waves may resonate with them, causing agitation.

We Are Tuned to Nature

Ever since the origin of life on earth, energies in nature have coconspired to support life. Electromagnetism is a property in nature, and it has existed ever since the origin of the universe. That has influenced and shaped evolutionary development of living species over the millennia. Evolution has shaped our senses to 'tune in' to the segments of the enormous range of energies as would benefit our survival. Apart from the senses of touch, taste, smell, hearing, and vision, our bodies interact in a subtle and intangible manner with electromagnetic fields in our environment.

With such an anarchic, enormously tangled, vertiginous web of so many kinds of waves all around us all the time, how on earth do we have an incredible ability to focus our vision around the object of our interest without being dizzy? This is so because in the multidimensional web of waves, our eyes are tuned in to wavelengths of visible portion between 400 and 700 nm, which are selectively sent to the brain as specific frequencies that our minds turn into colours that we see. For human eyes, leaves are green because leaves reflect the 510 nm wavelength of the colour green and absorb all other wavelengths of visible light, but we do not know how leaves reflect infrared or ultraviolet waves. Other segments of the spectrum contain information too, and that is a subject of scientific curiosity. Man has developed the ability to harness information in other segments of the spectrum as well.

In a dynamically interlocked system that we call a living body, there are interactive wheels within wheels. Cells form tissues, tissues form organs, and organs form the organism. For the organism to retain its integrity and to prevent its component systems from falling apart, a continuous system of internal as well as external communication is in place. It employs chemical and electrical works. Staying alive requires fast communication channels. Chemical communication alone would be too slow. This is why we can feel sensation of touch almost instantaneously. The coordination between hormones, chemical cellular processes, and the electromagnetic nature of the heart and brain is a continuous process that adjusts itself to the need of the moment.

In cellular communication, an electromagnetic frequency is transmitted before a chemical interaction occurs. Every cell acts as a small antenna. Brain cells are like liquid crystals and contain magnetite. All antennas, just as those in TV or radio, sense all radio signals but tune into ones that are of interest. The human nervous system is extremely sensitive to invisible signals from the planetary activity in the environment, though we may not feel it.

Disorders and diseases change not only the biochemistry of tissues, as detected by laboratory biomarkers, but also the electromagnetic properties of cells of tissues involved. The human body emits vibratory electromagnetic information that carries the signature of order or disorder. Biophotonics is a case in example. It is possible to diagnose disorders of the heart and brain by analyzing their electrical and magnetic activity via ECG, EEG and MCG, and MEG, respectively. Subtle energy systems are not supernatural. They go to foundations of life. The molecules and energy fields in our environment can affect living systems. Any pulsed signal (sound, light, electric, or magnetic wave) can entrain brain waves.

In the past, the mechanistic science has regarded the body's intrinsic electromagnetic phenomena as simply insignificant by-products of biochemical processes. Whether biochemistry drives them or they drive biochemistry is yet another question. The earth has its own natural electromagnetic background produced by itself and the cosmic source affecting living organisms.

A common explanation given for accuracy of science is that it is evidence based and that scientific evidence can be replicated. How was the evidence found? Research, of course. Have we conducted enough research in energy fields to prove or disprove the existence of alternative mechanisms of how the trio of body, mind, and soul might be at work?

We haven't. Why not? Who decides what to research upon?

Research needs funding. It is like an investment game. An investor would want to see tangible outcomes that bring return on investment. There never was enough money for research on the human biofield. The product of that research would have been very difficult to be put up on the shelf in a marketplace. Compare it with mechanistic research on biochemistry of the body. The more you unearth what happens at the molecular level, the more the opportunity to design a synthetic product that can influence the biochemistry and produce 'results' for everybody to see, no matter if those results are short-lived or may have adverse effects, which can always be downplayed. The synthetic product is then touted as a new wonder drug.

Bioelectricity is fundamental to life. That living systems can be affected by external electromagnetic fields makes perfect sense. As long as those fields are in nature, there is an excellent synergy between the two. The trouble begins when man-made fields upset delicate electromagnetic equilibrium between living species and the natural environment. The physical parameters of unnatural fields—such as intensity, pulsing pattern, and frequency—are unfamiliar for living species of all kinds. Every bit of ecosystem is under threat.

Chapter 3

Nature of Electromagnetism

> If you want to find the secrets of the universe, think in terms of energy, frequency and vibration.
>
> Nikola Tesla

Electromagnetism, other than its visible portion, can be counterintuitive and hard to grasp because it is beyond the scope of our senses. It is noiseless, invisible, and can't be smelled. It is like a wave of energy surrounding any electrical device. The higher the frequency, the higher the energy it carries. It is probably easier to conceptualize sound waves because they are mechanical waves that are audible to us. Mechanical waves travel through a medium (such as air, water, and even solid walls). Electromagnetic waves can travel through vacuum and any medium at the speed of light with few exceptions, such as Faraday cage. Predicting field strength at a given location with respect to a transmitting device or base station is a very complex issue. It depends on the original field, antenna characteristics, metallic and concrete obstacles, and simultaneous presence of fields from other sources.

Charge vs Current

Charge is a property of particles of matter to experience a force when placed in an electromagnetic field. Current, on the other hand, is point-to-point flow of charge across a circuit via electrons (charge carriers) initiated by a voltage at one end. Much like the pressure that drives water through a hose, voltage is the force that energizes and drives the charge. Once a voltage is applied to it, an electron does not have to physically move all the way to the end of the circuit to deliver the charge as that would make the process extremely slow. Instead it elbows the electron next to it, which in turn knocks on to the one next to it to complete the circuit.

Electrical switches are devices to either allow or interrupt the flow of charge. This is how light turns on and off instantly when a switch is flipped. Electric charge is present even when the switch is not yet flipped on. You do not pay utility bills for that. Electric current moves when charge becomes 'alive' as the flow begins, and the energy thus moved is used up, as in lighting a bulb, or transformed into another kind of energy, as turning on a heater or a microwave oven. That is what you pay for.

The difference between direct current (DC) and time-varying alternating current (AC) is that in the former current flows in one direction as in a battery, whereas the latter reverses direction every now and then—as in the power line supply in your home. As current begins to flow, a magnetic field begins to be set up instantaneously at right angles to the electric current. The two move together perpendicular to each other as well as to the direction of the flow of the current. When appliances are turned on simultaneously in our kitchens, a significant magnetic field may build up. The electromagnetic field around a direct current is stable and uniform, whereas the alternating current disappears and reappears with its poles reversed every time the current changes its direction.

The electric field and magnetic field are essentially two properties of the same energy. The magnetic field moves around the electric circuit when current flows. Depending on its strength at the source, the magnetic field from the source diminishes as it moves outwards. For instance, in case of residential wiring, it is about ¼ at 1 ft, and at 4 f, it is minimal. The electric field is measured as Volts/metre the and magnetic field as gauss in North America or tesla in Europe: 1 tesla (T) = 10000 gauss (G).

As most environmental EMFs produce tiny magnetic fields, we need smaller units to quantify them: 1G = 1,000 mG, and 1 T = 1,000,000 μT.

This is how to convert microtesla to milligauss: 1 μG = 10 mG, and 0.1 μT = 1 mG.

The quantum of irradiance of energy per unit area (power density or Pd) in an electromagnetic wave is the product of electric field strength (*E*) and the magnetic field strength (*H*). Thus, *Pd* (watts/metre^2) = *E* (volts/metre) x *H* (amperes/metre). In practice, it is more useful to use milliwatts and microwatts/centimetre^2 (see Appendix II).

Wireless Electricity

Before he died at the age of 36 in 1894, Heinrich Hertz had shown that electromagnetic waves travel through space at the speed of light. The electromagnetic field (EMF) is the field through which electromagnetic radiation (EMR) travels. As an electromagnetic wave propagates through space, energy is transferred at the receiving end. Electromagnetic energy, however, doesn't always have to be guided by a conducting wire. Nature is the foremost example of wireless technology. For instance, lightning during a thunderstorm and the arrival of sunlight at sunrise is an example of wireless flow of electromagnetic energy in nature. In the twentieth century, however, electromagnetic energy has been harnessed to a point that its flow across the space, unguided by any wire(s), has become subservient to man.

Passing alternating current through a wire that causes a magnetic field to oscillate around it instantly can artificially generate electromagnetic waves. Current flowing through the wire produces photons. If photons are in the frequency range of visible light, then emission of photons can be seen as light as in the case of the electric bulb. Travelling at 300,000,000 m/s, if an electromagnetic wave comes in contact with a metallic surface, it can induce current in it by inducing its electrons to get charged and cause flow of the charge in it. That is how the antenna was invented, which has made it possible to transmit and receive electromagnetic energy.

Antennas can decouple electromagnetic energy from the wires and allow it to be transmitted into space. The opposite is equally feasible. Their efficiency depends upon the material they are made of, physical configuration, and wavelength of electromagnetic energy to be transmitted or received.

Frequency and Wavelength

There are a handful of scientific terms with very specific meanings which should better be understood and done away with in order to navigate through the rest of this subject. Radiation is electric and magnetic energy coupled to each other at right angles, produced by the acceleration of charged particles propagating at the speed of light in all directions in the form of synchronized waves. A wave typically has an upward crest and a downward trough before it returns to neural point to complete one cycle. The number of such cycles per second is called its frequency and described as hertz; 1,000 hertz (Hz) is 1 kilohertz (kHz), 1,000 kilohertz (kHz) is 1 megahertz (MHz), and 1,000 meghertz (MHz) is 1 gigahertz (GHz). The distance between two crests is called its wavelength. Wavelengths vary from thousands of kilometres to nanometres. For instance, the wavelength of visible light is akin to the size of bacteria, radio waves that bring us radio and television signals are the length of cars or buildings, gamma rays have a length no more than atoms.

Frequency and wavelength are inversely related, and their product is equal to the speed of light. The electromagnetic spectrum is comprised of a vast range of frequencies. As we shall see later, the properties of electromagnetic waves change as the frequency changes. Magnetic fields are produced when current actually flows. The coupling of electrical and magnetic components and therefore the intensity of radiation is stronger as the frequency increases. Power lines produce magnetic fields continuously because current is always flowing through them. In residential supplies at 50 or 60 Hz, the electric and magnetic components have a very weak coupling. Therefore, electric and magnetic field strengths must be measured separately.

The length of each electromagnetic wave described as its wavelength has an inverse relationship with its frequency. Thus, a 1 Hz wave is about 300,000 km long, a 10 Hz wave is 30,000 km, 50 Hz wave is 6,000 km,

60 Hz wave is 5,000 km, a 1,000 Hz wave is 300 km, a 1,000 kHz or 1 MHz (such as in AM radio transmission) is 300 metres, a 100 MHz wave (such as in FM radio) is 3 metres, a 1,000 MHz (1GHz) wave is 30 cm, and a 10 GHz wave (as used by satellites) is only 3 cm, a 300 GHz is 1 mm, and 3000 GHz at 0.1 mm (see Appendix I).

Electromagnetic Spectrum

The types of electromagnetic radiation are broadly classified into the following classes:

- ionizing radiation
    1. gamma radiation
    2. X-ray radiation
    3. ultraviolet radiation.
- nonionizing radiation
    1. visible radiation
    2. infrared radiation
    3. terahertz radiation
    4. microwave (radio frequency)radiation
    5. radio waves
    6. very low frequency radiation
    7. extremely low frequency radiation.

This classification goes in the increasing order of wavelength, which is characteristic of the type of radiation. The key point is, except for visible light and infrared EMF, we cannot perceive any of these energies without instruments designed to detect them. However, our brain senses them all without us being conscious of the sensation. This is why we are unaware of the impact of abruptly added massive man-made EMFs to our environment.

ELECTROMAGNETIC SPECTRUM

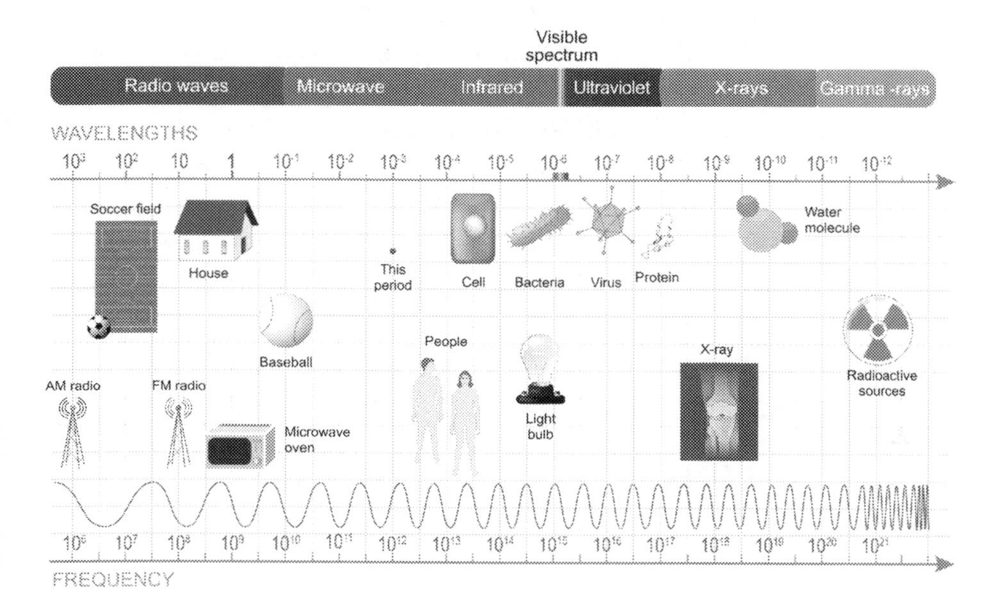

Fig. 1 Electromagnetic spectrum.

Based on frequencies, the International Telecommunications Union (ITU) has classified nonionizing electromagnetic waves into 12 bands from extremely low frequency (ELF) EMF to tremendously high frequency (THF) EMF. Though in real life some degree of overlap may occur, the classification conveniently facilitates descriptions and is indispensable in the allocation of spectrum for devices. Quite like the keys of a piano, different segments of EMF spectrum behave differently.

Man-made electromagnetic noise is generated from technologies for power transmission and communications and is generally concentrated in particular frequencies, especially in urbanized areas.

In the ELF band (3–3,000 Hz), the man-made source is power lines, which generally operate at 50 Hz (60 Hz in USA).

Power frequency electromagnetic fields are generated in the production, transmission, and distribution of electricity. Residential and commercial electric supply is through these frequencies. They couldn't have been wireless because their wavelengths are so large that, in order to propagate them wirelessly, one would have to build extremely large antennas, which is impractical except in certain limited circumstances (such as communication with submarines), as these frequencies can travel underwater. From the generation plant, high-voltage power is stepped down as it transmits to homes via substations.

Transmission lines are either overhead or underground. Overhead lines produce both electric and magnetic fields. Electric fields of underground lines remain under the surface of the earth but magnetic fields can be observed above the surface as well. These ground currents (10–20 mG) have been known to affect cattle. Cows do not give as much milk. In case of human residential colonies in close proximity to power lines, childhood leukaemia has been related to ELF EMFs.

Power lines, wiring, and electrical appliances, such as electric shavers, hair dryers, computers, televisions, and electric blankets produce what are called extremely low frequency (ELF) EMFs. Most of our electronic devices (such as kitchen appliances, PC monitors, TV sets, AM radio, etc.) operate in very low to high frequency band. In the very-high to ultra-high band, emissions come from FM radio, TV, mobile phone base stations, automotive ignition, cordless phones, and microwave ovens.

Satellite systems, radar, scientific apparatuses for astronomical observations operate in super-high to extremely high bands of nonionizing radio frequencies. Radio broadcast used to be in three ranges—namely, AM (amplitude modulation), FM (frequency modulation), and SW (short wave). AM is now obsolete. FM in Band II is now the standard. SW still has a place because its high frequency signals can be reflected from the ionosphere to go across the globe. Band III was standard for TV broadcast, but that has now been elevated to Band IV and IV. Health risks from radio and television antennas were not controversial for a long time because they were few and far away from residential areas and were not in as large numbers as mobile phone masts.

Ionizing vs Nonionizing Radiation

Distinction must be drawn at this point between ionizing and nonionizing radiation, especially with regard to their interaction with biological systems. Ionizing radiation has the ability to damage components of cells. On the other hand, nonionizing radiation, such as radio waves and microwaves, do not have similar capabilities and can at best cause heating of tissues. This claim though has been seriously doubted after a thorough scientific scrutiny, as we shall see later.

Electric fields are easily shielded or weakened by walls and other objects, whereas magnetic fields can pass through buildings, living things, and most other materials. Consequently, magnetic fields are the component of ELF EMFs that are more relevant with regard to health. With a 50 Hz wave having a length of 6,000 km, the impact is vastly spread out compared to the size of a human body such that more than 3,000 people can stand with their arms stretched out within the perimeter of just one wave, so the net impact is small. On the other hand, with an X-ray having a wavelength of a billionth of a metre, at least one-third of a billion waves can pass through one adult human being in one single exposure, and the net impact is huge. For high frequency ionizing radiation (such as X-rays, gamma rays, and cosmic radiation), the quantum energy of radiation itself can break the bonds between molecules. This is why they are classified as ionizing radiation.

Natural vs Man-Made Sources

One of the major differences in natural versus man-made sources is that the former is non-polarized, whereas the latter is polarized. This is important because this characteristic is extremely relevant to the interaction between electromagnetic noise and biological tissues, apart from other characteristics (such as power, frequency, continuous versus pulsed, and modulation). Man-made electromagnetic waves can be generated by oscillation circuits through induction of oscillations of electric charge in one direction. This is why they are called linearly polarized.

Natural fields are mostly DC (direct current) static fields produced by oscillation of atoms or molecules in all directions and are thus non-polarized. Cells in our bodies recognize them since the beginning of evolution. If you use polarized sunglasses while driving, non-polarized light is converted into polarized light, thus removing glare from oncoming traffic. Polarized waves are more likely to interfere with organic material, such as biological tissues, as the latter contain charged polar molecules. The interference occurs in the form of inducing forced vibrations in the charged/polar molecules of living tissues. Schumann resonances are at 7.83 Hz. The human heart beats are at 1–1.5 Hz. The human brain vibrates at 7.8 Hz when in harmony. Mobile phone electromagnetic emission is polarized and pulsed and at millions of hertz. In people who have photosensitive epilepsy, a light flashing at a frequency as low as 15 Hz can induce an attack. Several children were reported to have experienced it when ninja video games were first introduced.

Fields from man-made and natural sources of RF fields vary in their spectral and time domain characteristics. This complicates the comparison of their relative strengths. For example, if you consider a bandwidth of 1 MHz, man-made fields are stronger than background natural ones, whereas if you consider the entire range of 300 GHz, natural fields may be stronger than man-made ones. While estimating exposure from an emitter (e.g. a mobile phone), there are several considerations, such as frequency of incident wave, distance from the emitter, angular direction, and environmental modifiers such as reflection, diffraction, shielding, etc. All these can modify the fields. By the time the specific energy gets absorbed into body tissues, the actual dose is fairly reduced.

In a nutshell, there are three sets of electromagnetic frequencies that have biological effects. At the lowest end of the spectrum, we have ELFs (such as power line EMFs), which affect living organisms by magnetic influence.

At the other extreme, we have X-rays and gamma rays with ionizing properties having the ability to damage components of cells, tissues, and enzymes by damaging DNA, resulting in mutations in protein lines and the breaking of chemical bonds affecting the natural apoptotic chains and interfering with the intercellular signalling cascades by which cells communicate and divide. There is no minimal threshold for affecting these changes. Indeed, certain medical uses, such as diagnostic computerized tomography and radiation for cancer treatment, are due to this property of electromagnetic radiation in the range of X-ray and gamma ray frequencies.

Somewhere in the middle, we have nonionizing EMFs, which cannot break chemical bonds between tissues and can at best cause heating of tissues. That has been the traditional view. However, the man-made radio frequencies make a very special case. This is so because in the latter, there is a carrier radio frequency wave, typically 1,800–2,400 GHz, and pulse-modulated over it is the secondary information wave, which carries audio, text, and video data packages. These are called radio frequency electromagnetic fields (RF EMFs) or microwaves.

When magnetic resonance imaging (MRI) was introduced to medical diagnostic armamentarium, it was touted as completely safe as it did not emit ionizing radiation unlike computerized tomographic (CT) scanning. That they are completely safe is a lie. In MRI hydrogen atoms in body fluids are made to resonate in an extremely strong magnetic field at 1–3 T, but sometimes it's up to 8 T such that they emit radio frequency radiation at the resonant frequency. Variations in water content are the basis of contrast in the images obtained. In response to applied radio frequency signal, which is accompanied by magnetic field, protons of hydrogen atoms of water transfer from a lower energy state to a higher energy state to resonate with the applied signal. When the protons fall back to a lower energy state, RF is emitted and picked up as an image.

Surgical diathermy employs low RF fields at 500 kHz with harmonics up to 20 kHz. Robotic navigation surgery nowadays requires the EMF generator to be placed right under the head as a replacement of regular headrest. Induction cooking hobs have coils that produce a magnetic field below the metal cooking pans to heat them. Microwave ovens operate at 2.45 GHz and a power setting of 500 W to 2,000 W.

A whole lot of newer technologies based on radio frequency in our daily lives include smart metres, air traffic control, marine radar, tracking radar, medical diathermy, whole-body security scanners at the airports, antitheft security devices, and radar systems used for monitoring weather, overspeeding on highways, vehicular collision avoidance, toll tax monitoring, and tracking RFIDs on household appliances.

Carrier vs Information Wave

An electromagnetic wave with no information on it is a carrier wave. For it to carry information (such as audio, text, or video), it has to be modulated; that is, an audio, text, or video signal has to be 'mounted' on it. It is not like a person carrying a load. It is rather like a carrier wave being energized by information. They have zero mass. This is called modulation.

At the receiving end, that signal has to be demodulated once again to become an audio, text, or video signal. The receiving antenna has a built-in tuner that allows the receiver to tune in to the intended frequencies only. This is typically done via a resonator—a capacitor and an inductor creating a tuned circuit—which amplifies oscillations strictly within a certain frequency band and attenuates other frequencies that might have been picked up by the receiving antenna.

The distance over which a transmitting antenna and a receiving antenna can do a tango together depends on the power of the transmitter; quality of the receiver; size, shape, type, and height of the antenna; mode of transmission; and interference by other signals.

When televisions used an analogue signal, they had to be tuned manually. Newer televisions tune themselves automatically. Modulation essentially implies that an information signal is mounted on a carrier wave and either its amplitude or frequency gets changed. For instance, AM radio utilizes amplitude modulation, and FM radio utilizes frequency modulation. AM broadcast has fallen out of favour because its audio output sounds scratchy as its carrier wave frequency is limited to 5 kHz, whereas human hearing occurs between 20 hertz and 20 kHz. FM radio sounds better as its frequency output is up to 15 kHz.

Your wireless modem does much the same with multiple carrier waves although via far more complicated arrangement called orthogonal frequency division multiplex (OFDM). Technologies for modulation in the digital era have evolved to a great degree. Unlike the dots and dashes of Morse code of telegraphy days, digital systems use 0s and 1s in various combinations as electromagnetic equivalents of information signal.

Every device that connects to wireless network does so through a Wi-Fi card built into it that sends and receives microwave signals. A wireless router is a sort of access point that mediates between your laptop, PC, iPad, or smartphone and the network of your service provider. The essential component of a router is a modulator and demodulator (modem) device.

Heart of the Controversy

Whereas biological tissues would not respond to a carrier microwave except through heating, depending on its wattage, they do respond to the secondary information wave, which carries the information (such as audio, video, or text). *This* is what lies at the *heart* of the entire controversy.

It wasn't clearly understood at the time when mobile phones were first introduced in the open market in the mid eighties of the last century. The mobile phones operate at a very low power at about 1 W or even less. The microwave ovens for domestic use operate at 750–1,000 watts. Because the RF waves used in the mobile phones would neither heat up tissues nor ionize chemical bonds at molecular level, it went without saying that they must be safe for human use. Besides, a mobile phone wasn't, after all, a pill that you would swallow. So no big deal about safety.

Indeed, there were about 15 million users in USA alone until the 1990s without any premarket safety studies ever having been conducted. And that being so, there was no question of postmarketing surveillance either. In 1993 on a *Larry King Live* television show, David Reynard, whose pregnant wife had developed a brain tumour after using a Motorola mobile phone, brought attention to the brain scans of his wife after her neurosurgeon was convinced that the tumour depicted on her scans was caused by a mobile phone as suggested by its location, morphology, and direction of spread from outside in. The story went viral.

Motorola stocks took a dip. The Cellular Telephone Industry Association (CTIA) addressed the media in a press conference to allay fears in public mind, citing studies that showed radio frequency microwaves were harmless. The rationale was evolved by comparing microwave ovens with mobile phones. Ovens could heat food, but mobile phones didn't heat the body. It was named low-power exclusion.

It was later alleged that they had shown studies related to microwave ovens because there were no studies conducted on cellular phones up until then! Heads rolled, and the government turned to Food and Drug Administration (FDA) to find the truth. An urgent need for conducting research on the subject was felt, and an onus to raise money for the same was put on the industry. CTIA agreed to fund $28.5 million for research on the health risks of radio frequency microwaves, provided the government wouldn't regulate them until results of studies were available. Deal done. It was later learnt that some members of FDA had shares in AT&T, the major player in the wireless industry in the USA at the time.

George Carlo, who held a degree both in science as well as law, was entrusted with the gigantic task that would get under way in 1994 for next 5 years. Over 200 scientists were involved in several centres to conduct 56 epidemiological in vivo and in vitro animal studies, which were peer-reviewed by Harvard School of Public Health. When his results turned out to be contrary to the expectations of the industry, Tom Wheeler, the then chief of CITA, who represented the interests of the industry, called up George Carlo, thanked him for his good work, and dismissed the issue with the remark that the findings were inconclusive. Recently, Tom Wheeler was appointed chief of Federal Communications Commission (FCC), the governmental agency that oversees regulation of the wireless industry in the USA.

When they warn you to keep your mobile phone at least 15 mm away from the body, you might be duped into believing that it is intended to guarantee your safety. The truth is far from it. If you got brain cancer and sued the mobile phone company, lawyers will ask you to prove that you kept your Apple iPhone at least 15 mm (25 mm for Nokia or Motorola) away from your body over a long period of use. Even if you won the suit, the insurance companies would refuse to entertain claims for health risks emerging from use of mobile phones, which is a standard exclusion in all insurance policies.

Safely concealed in the tiniest-sized letters, the safety information of mobile phones says, 'This device meets RF safety standards when the device is used at the specified distance from the body.' In other words, they know as well that the device is not safe if worn on the body, as most of us do.

Chapter 4

Grandfather of the Mobile Phone

Though World War I was fought in the twentieth century, the means employed were still steeped in traditions of previous centuries, with infantries engaging in pitch battles, employing rifles, machine guns, poison gas, and even barbed wire.

While the world was recovering from the horrors of World War I, little was known to the rest of the world that somewhere a fierce leader was emerging, who dedicated the first six years of his chancellorship of the Third Reich in developing a secret and unparalleled underground war machine that would be unleashed in 1939 with the intent of ruling the world for a thousand years. His name was Adolf Hitler.

Wars Inspire Innovation

In the period between the two great wars of the twentieth century, the scientific world was agog with ideas about changing the face of warfare technology. The need to have an edge over the enemy has been a strong determinant in advancing military technology throughout human history. No wonder then that during World War II, some eminent scientists have had to be bedfellows with warmongering politicians due to expediency of the moment. Marvels of technology that get passed on to civilian use have, by far, had their origin in research dedicated to military use. But for the monstrous challenge posed by Adolf Hitler, several technological advances in the early part of the twentieth century wouldn't have occurred in such a hurry as they actually did.

A certain degree of knowledge about electromagnetic waves had been gained from discoveries made in the latter half of the nineteenth century by Maxwell and Heinrich Hertz. The time to convert them into technology had arrived.

In the little-known village of Litchborough in the south Northamptonshire, England, stands a modest memorial bearing names of two men, Robert Watson-Watt and Arnold Wilkins. In February 1935, they drove a van to Litchborough, carrying a set of wooden poles and a radio receiver. Using a hired BBC short-wave radio signal, they demonstrated that the signal was reflected back by a Heyford bomber flying 8 mi. away. The experiment was to be the key to win the Battle of Britain five years later.

The Committee for the Scientific Survey of Air Defense (CSSAD) headed by Henry Tizard took a serious note of Watson-Watt's idea and encouraged them to develop it further. They developed a system to detect the reflection of radio signals from metal objects several miles away; it was named radio direction and finding. Within a year, five stations surrounding London were erected. Within the next two years, a ring of 350 ft high towers of early warning systems was built along the eastern coast of Britain from Portsmouth in the south to Scotland in the north to create an air shield against enemy aircrafts flying anywhere within 100 mi. off the coast; it was known as the Chain Home system.

On 14 September 1938, the German airship LZ 130 Graf Zeppelin flew as the first-ever electronic surveillance mission, but it failed to figure out the Chain Home system. Perhaps it used the wrong frequencies. Nevertheless, it wasn't until the beginning of the war that the system was put to a real test. A radar system transmits radio waves in the microwave range of electromagnetic spectrum that can be partially reflected from metallic objects and give an estimate of distance, altitude, and speed of an object such as an aircraft.

Although functional, the Chain Home system was far from perfect. In February 1940, John Randall and Henry Boot of the University of Birmingham developed resonant-cavity magnetron; it could produce a microwave beam in a much higher frequency range and much shorter wavelength, which was a huge upgrade in radio direction and finding.

Had that happened five years earlier, Britain would have been in a far better situation to industrialize production of the new version of radar on a mass scale.

With the uncertain political climate of the times, Churchill apprehended the imminent invasion by Germans. With a stroke of political sagacity, he agreed to trade the secret of the magnetron with the USA in exchange for a speedy supply of stocks of radars. On 31 August 1940, Henry Tizard led the mission to meet Vannevar Bush, then head of National Defense Research Committee (NDRC) of the USA. A demonstration of the magnetron by the British team at Bell Labs had American jaws dropping in awe and wonder. In quick succession, a radiation laboratory was set up at Massachusetts Institute of Technology, and within months, Raytheon Corporation had started bulk production of sophisticated radars.

Encouraged by a string of easy victories over Czechoslovakia, Poland, Norway, Holland, Belgium, and France, Adolf Hitler had ambitions to rule the world. The problem was a tiny island that stood across the narrow English Channel—Britain. It wasn't so tiny, after all, as Britain was the leader of the British Commonwealth, the global superpower of the time. It was time to avenge the defeat of World War I. In the summer and fall of 1940, the skies over United Kingdom were lit with the first-ever all-air bombing campaign, and the British Royal Air Force, with its air power only one-fourth that of Germans, thwarted all attempts of the German Luftwaffe to gain superiority. Adolf Hitler gave up his plans of amphibious invasion of Britain through Operation Sea Lion. By the end of September 1940, the Battle of Britain was over. But it had given the world a very powerful weapon for all future wars—radar!

Microwave Oven

Percy Spencer, an engineer from Maine, USA, worked for Raytheon Corporation on a mission to improve the magnetron technology. On an afternoon, while working in his lab in front of a microwave beam from magnetron, Spencer was hungry. He reached in his pocket to get his candy bar but was horrified to find it had melted.

Instinctively, he knew it had to be the microwave beam from the magnetron. Next he placed corn in the path of the beam in the presence of colleagues. Voila! Corn popped. Unable to hold their curiosity, next they tried to scramble an egg. Voila again! It scrambled so well that it splattered in the face of one of their colleagues! Subsequently, Spencer created a high-power electromagnetic microwave beam from a magnetron, placed it inside a metal box to prevent it from escaping, and when food was placed inside the box, it became warm. Microwave oven was born—thanks to radar!

In 1947 sumptuous hot dogs were served from a 5 ft 11 in. microwave oven, then called radar range, costing US$5,000 (US$55,000 in today's value) at Grand Central Terminal, New York, for the first time in the world. Today tabletop microwave ovens are at $50 apiece and use an electromagnetic beam of 2.45 GHz and a wavelength of 12.2 cm to heat up and cook food by a process called dielectric heating.

The journey of radio wave transmission from telegraphy, radio, television, fiber optics and satellites to mobile phones and the Internet is fascinating.

If radar is the *father of the microwave oven*, then it truly is the *grandfather of the mobile phone*!

Every invention is preceded by another invention. To entirely credit one person for a particular invention would be unfair. Despite this, there is a tradition to refer to inventions after names. Though Nikola Tesla had attempted to patent the invention of radio transmission, it was Guglielmo Marconi who first demonstrated the use of electromagnetic radio frequency waves in transatlantic radio transmission towards the end of the nineteenth century. But then much before that, it was James Clark Maxwell who had predicted such waves on the basis of mathematical calculations, without which Heinrich Hertz couldn't have actually demonstrated their existence.

Radiophones had been around a long time for policing in the airline and security industries. Indeed, since the use of the landline in 1877 by Alexander Graham Bell, even before the display of the first electric bulb in 1886, AT&T virtually ruled the business of telephony for almost a century. It included its second invention, the wireless car telephone, from the 1940s. The number of subscribers of landline phones took 90 years to reach a figure of 100 million users. AT&T had predicted that users of car phones would reach a figure of close to a billion. It wasn't destined to happen.

In comes the seductive *handheld, wire-free* cellular phone!

If there were no Maxwell or Hertz, there couldn't have been Martin Cooper. Working for a then relatively small company named Motorola, on 3 April 1973, he made a dazzling display of the first-ever handheld, wireless cellular phone DynaTAC 8000x in a street in Manhattan, New York, for passers-by and stole the glory from its rival corporate giant, AT&T and its Bell Telephone Company, even before walking into the press conference at the Hotel Hilton.

There were no integrated circuits (ICs) available at the time. DynaTAC weighed 1.1 kg because it had a number of conductors and transistors crammed in that consumed a lot of power, draining its battery in twenty minutes. You couldn't have carried it in your pocket. It was nicknamed the brick phone and even a shoe phone! *Popular Science*, an American magazine, carried the picture of DynaTAC on the cover of its July 1973 edition. Unlike a landline phone, the subscribers of the mobile phone reached a figure of 100 million in just 17 years (1983 to 2000).

The car phones disappeared very soon. The landline phone is fast becoming a pale shadow of its old glory. The story of the rivalry between AT&T and Motorola is a story of drama, conflict, and suspense.

Two-way radio communication using a particular band of frequency is a concept that has been used in creating an enormous range of technologies. In order to maintain fidelity between transmitting and receiving units at two ends of communication, a specific bandwidth has to be used for every application. Or else the technology will not work because far too many frequencies will interfere with one another. For this reason, before a certain application can be marketed, it has to have its specific frequency authorized to it. Even the design of the application cannot be made unless it is known beforehand on which frequency it will operate. The range of frequencies available for use by wireless technology is between 3 kHz to 300 GHz.

In the wee hours of 15 April 1912, RMS *Titanic*, a British passenger liner on its maiden voyage from Southampton to New York, carrying some of the wealthiest people of the time, collided with an iceberg in the Atlantic Ocean. Guglielmo Marconi's wireless telegraphy system, installed aboard RMS *Titanic*, with its antenna nearly as long as the ship, is credited for the survival of over 700 lives out of 2,200. Had they depended on flags or carrier pigeons for communication, no one would have survived.

When Marconi demonstrated the use of wireless communication, his equipment was not competing with any other radio frequencies. He was free to use any part of the spectrum. But now that there are millions of competitors vying to use a slice of the pie, it needs to be strictly regulated.

Every licensee's signals have to be protected now more than ever because of problems of hacking and loss of privacy. In the 1970s, fierce legal battles were fought between radio common carriers and independent telephone companies—especially Motorola on one hand and the giant of nearly half a century until then, the Bell Telephone Company of AT&T, on the other—for a share in the pie of frequencies to be allocated by the Federal Communications Commission in the USA.

The introduction of wireless broadband and rapid expansion of 4G and 5G has further increased demand for spectrum allocation. Currently, it is done by an auction method to increase competition in evolving more innovative technologies. RF spectrum—like land, minerals, and water—is now seen as a national resource and therefore a property of the state. It is necessary to avoid interference of frequencies between different applications, and therefore generation and transmission of radio waves are strictly regulated by national laws and coordinated by the international body International Telecommunications Union (ITU).

With lower frequencies, providers could cover a large area. With higher frequencies, providers could provide to more customers over a small area. In cellular network design, each cell site or tower provides connectivity in a small geographic area called a cell, which is connected in a contiguous manner to provide seamless coverage. The radio frequencies allocated to a particular provider used in a cell can be reused in a cell located far enough to eliminate interference. Thus, a single provider can cover a very large area.

Mobile phones use a duplex two-way link to send a message to a base station (uplink) or receive messages from a base station (downlink).

The number of transmitting base stations required depends on the number of users. In densely populated areas, they may be required just as far apart as 100 metres.

When first introduced in 1980s as 1G, these were analogue systems, such as AMPS (Advanced Mobile Phone System) in the USA, TACS (Total Access Communication System) in Europe, NMT (Nordic Mobile Telephony) in Sweden and Scandinavia, and NTT (Nippon Telegraph and Telephone) in Japan. They functioned through 1990s and were slowly phased out well before 2000. Introduced in 1990s, 2G were based on digital protocols. There were two main types: TDMA (time division multiple access) and CDMA (code division multiple access). TDMA subtypes are GSM in Europe and Asia, PDS (personal digital cellular) in Japan, and D-AMPS in North America.

3G was the earlier prototype for today's smartphones, a technology finally realized with the advent of 4G. Some of these are based on LTE (Long Term Evolution), which uses OFDM (orthogonal frequency division multiplexing) protocol and others on WiMax (worldwide interoperability for microwave access).

With analogue systems, frequency bands used for cell phones were below 900 MHz. With GSM and TDMA for 2G, frequency bands used were from 890 MHz to 1,910 MHz and, with CDMA, up to 1,980 MHz. With LTE and WiMax, many more are possible!

A mobile phone is essentially a two-way radio communication device that uses radio frequency microwaves for transmission. These are the same waves with which we communicate with satellites or cook popcorn in a microwave oven. With old models, the antennae of cellular phones were designed in such a way that they proudly jutted out of the handset almost as if to convince you that they were not just toys.

When awareness about the effects of RF radiation became known, the antennas disappeared. Well, they didn't actually disappear. They are well concealed inside the handset. It is important to know where they are located because that is the spot where radiation plume is most intense.

In standby mode, transmissions between the handset and base station are brief and infrequent, but if your handset is equipped with data service, such as emails, then transmission may be more intense as the mobile phone tries to contact remote servers. Local emission during data download is higher than during calls. No wonder they are called 'smartphones'.

In texting, transmission is very quick, typically less than 1.5 seconds. The earpiece is thought to be safe, but some doubt has been raised about that. Bluetooth would reduce intensity of exposure, but because two devices would be constantly searching each other, the exposure is prolonged.

4G models have built-in safety technology, such as DTX for adaptive power control, which means these phones continuously adapt their radiated power to minimum necessary for satisfactory communication. Older cordless phones didn't have that. Newer DECT phones operate in 1,880–1,900 MHz range and may have adaptive power control. The signal between its base and handset is discontinuous. Adaptive power control was designed in the hope that it would reduce biological impact, but in fact, that didn't happen.

The frequency bandwidth required for the new smartphones is greater than before. In order to achieve this, they have developed 'smart antennas'—a series of four antennas in a single handset, which is equivalent to having four mobile phones in one. The broadband infrastructure of base stations to support this is equally being expanded.

Mobile phones emit RF waves to nearby cell towers, which are like antennas. So are our bodies act as antennas and absorb the RF energy, converting it into alternating eddy currents. The transmitter in a mobile phone operates at 0.75 to 1 W of power with a maximum of 2 W at peak usage. When in talk mode, it also creates a radio frequency plume around itself. In pulsed mode, electric current moves back and forth. The plume is least in standby mode, more when it is ringing, but much more when you speak into it than when you listen to it.

In the 1980s, the first cell phone operated in analogue fashion at 902.5 MHz (NMT). A decade later, GSM replaced it, transmitting at 902.4 MHz, pulsing at 217 Hz. The current digital communication system (DCS) operates at 1,800–2,400 MHz. The regulatory body for radio frequency EMF in USA is ANSI, which is a part of IEEE along with FCC and USEPA. In the UK, it is NRPB. These bodies are supposed to give information on current regulations and represent the government on health concerns.

Chapter 5

The SAR Hoax

The International Association of Research on Cancer (IARC) includes nonionizing radio frequency of 30 kHz (wavelength 10 km) to 300 GHz (wavelength 1 mm) as the part of electromagnetic spectrum that is a matter of concern for public health. Radio waves with frequencies between 30 kHz to 300 GHz are nonionizing microwaves though it does not mean that the abrupt changes in their physical properties are sharply defined by these arbitrary cut-offs.

Exposure to this range of frequencies can occur in one of following ways:

1. environmental: mobile phone base stations, broadcast antennae, smart metres, medical applications, such as MRI, hyperthermia.
2. occupational: high frequency dielectric and induction heaters (cooking hobs), remote detection of objects and devices (antitheft devices, radar, RFID), high-power radars.
3. personal: mobile phones, cordless phones, laptops, Bluetooth devices, RFID, iPad, Google Glass, amateur radio, radio, TV, and wireless networks.

In the United States, the legal position adopted by US FDA and FCC is looked up to as reference for safety standards in telecommunication consumer products. In Europe, standards are set by ICNIRP. Other countries follow one of these standards.

SAR

Specific absorption rate (SAR) is a technical specification of a device that uses RF EMFs and indicates the rate at which RF energy is absorbed by the human body when exposed. It is defined as the power (in watts) absorbed per unit mass of tissue (in kilograms). Calculations of SAR involve measurement of the loss of electric energy from the incident RF wave, electric field strength induced in the tissue, and the rise of temperature.

Over a period of six minutes, two metrics are determined—namely, wbSAR (whole-body SAR), which is the total EMF power absorbed by a body divided by its mass, and psSAR (peak spatial SAR), which is averaged over any cube inside the body with a tissue mass of either 1 g (psSAR-1g) or 10 g (psSAR-10g).

The value (Watts/kg) is used to set standards for safety of use of the device, supposedly from the perspective of health effects. Though it puts a limit on maximum permissible levels of exposure that a device such as a mobile phone must not exceed, the complexities involved in trusting this specification as a measure of safety are immense. In the United States, FCC recommends 1.6 W/kg averaged over 1 g of brain tissue as a safe peak spatial SAR limit for exposure of the head. For the whole body, the limit is 0.08 W/kg, whereas the accepted limit for exposure to hands, feet, wrists, and ankles is 4 W/kg. In Europe, ICNIRP recommends 2 W/kg averaged over 10 g of brain tissue.

SAR meausrements are made using an active mobile phone placed close to a model of a human head. The model is called specific anthropomorphic mannequin (SAM) and supposedly represents an average human head. Well, in fact, it represents an average size of the head of a man six feet two inches tall and weighing over 200 lb. The SAR for a 10-year-old is 153% more than the SAR for the SAM model.[1] Spare a thought for babies exposed to RF radiation in baby monitors and schoolchildren exposed to Wi-Fi in classrooms!

SAM is filled with a fluid which supposedly simulates the electrical properties of human tissues. A robotic probe is then placed inside the model head to measure electric fields. The experiment is done through every frequency band of cell phone at its maximum output of power while changing the positions of mobile phones, and SAR value is then calculated by averaging several measurements.

It is plausible to assume that it is only the 'absorbed' part of EMF radiation that can induce interaction between EMF and certain biological molecules. Thus, SAR should be a key measure. The technique of measurement of SAR basically looks for conversion of absorbed radiation from RF EMF into thermal energy. However, that is irrelevant to living species because damage has been seen at levels much below that. The problem, therefore, is to define what is non-thermal effect and how to distinguish between direct and indirect thermal effect.

It Is a Hoax

Some simple arguments would deflate the SAR logic used by industry to defend itself. To begin with, SAR calculations were made in the mid 1990s and are still used as such, whereas the frequency characteristics of cellular phones have gone a great distance since. SAR refers only to radiation arising from mobile phones. Today we are exposed even if we are not using a mobile phone. In 1996 mobile phones were so big that they were nicknamed brick phones or shoe phones. No one carried them in their pockets. Now when people carry them on their bodies, the exposure characteristics have obviously increased a great deal.

Besides, SAR inside the brain depends on the position of the mobile phone antenna relative to the human head, distance from the body, anatomy of the head, age of the user, quality of connection between the handset and the base station, level of battery charge, and whether you are in an open space or a closed space, such as car. It also varies when you are sending an audio message and when you are receiving one.

It is likely that the invasive probes perturb the radiation at the tip of the measuring probe and thus insert inaccuracy into the measurement of absorption of radiation. Also unlike in the head of SAM, tissues in the brain are not homogenous, and rates of absorption vary at different areas. The parameters used to determine SAR values in the laboratory do not represent actual exposure characteristics in real life.

Besides, laboratory technique does not take into account the fact that, in the case of children, the skull is not as thick as in an adult and radiation absorption is likely to be much deeper.

Average exposure from the use of the same mobile phone is higher by a factor of 2 in a child's brain compared to an adult and is higher by a factor of 10 in the bone marrow of the child's skull. The rise of temperature inside a human brain from use of a 3G phone is approximately 0.1 °C. The use of a hands-free kit lowers the exposure to less than 10% of the value from use at the ear but may increase exposure to other parts of the body.

Exposure from TV and radio antennae is much lower than base stations. Besides, because of frequency characteristics of RF waves used for TV and radio broadcast, too many base station antennae are not required. That is the reason that the pitch of revolt against MW technology was not so high when TV and radio were introduced long before mobile phones. Occupational exposure as from military radars and ATCs at airports causes long-term cumulative deposition of RF energy in the body much greater than the mobile phones, but spatial peak SAR in the head is less. Do not *barbecue* the chicken. *Boil* him on slow fire.

Tissue heating is a definitely established mechanism of biological tissue damage. Its absence in case of mobile phones due to very low wattage has served as an alibi for the industry for far too long. In the last decade and a half, however, several cellular molecular mechanisms that lead to damage at subthermal levels have been demonstrated. The most widely studied technology is the mobile phone for obvious reasons.

Non-thermal effects obviously cannot be as robustly tangible as thermal effects. However, there is a consistent pattern between the use of mobile phones and certain health problems. Intangible effects at cellular level are of great importance in setting limits to the design of such technology. The assumption underlying the application of SAR as a standard of safety is fundamentally flawed. Bodily functions are electrochemical. Nerve impulses are in the range of millivolts and brain waves in microvolts. SAR (Watts/kg) is supposed to indicate the level of absorption in the body tissues, but it actually refers to 'heating effect' and not the biochemical effect of radiation.

Changes in brainwave activity can occur with EMF exposure as little as 0.1 W/kg SAR. Besides, the way SARs were arrived at didn't take into account varying densities and conductivities of biological tissues, as all components of tissues do not respond to RF EMFs in the same fashion. Interaction between polarized electromagnetic wave and living matter occurs when an incident wave causes forced vibrations in the charged/polar molecules of living matter, thus transferring a bit of its energy. In biological tissues, energy absorption is not uniform in all tissues. The extent of damage produced depends on the structure and function of the tissues involved (e.g. the same amount of energy absorbed by DNA is likely to produce far more damage to the organism than if absorbed by lipids).

The degree of oscillations produced in living matter by natural electromagnetic radiation is millions of times smaller and indeed so subtle that it is organized by nature to nourish the living matter. Or else how would we produce vitamin D in our bodies via exposure to sunlight?

Allan Frey[2] reported that the microwave hearing effect could occur with modulated radio frequency even in deaf people several hundred feet away from transmitting antenna at as low as 400 $\mu W/cm^2$. The hearing took place not through the ear but was directly perceived by the brain. Therefore, even in the experiments done, the real absorbed dose and the delivered dose remain unknown.

It is ridiculous to think of biological tissues as electrically passive substances and conduct studies on it purely based on principles of electrical engineering. SAR should, in theory, be more reliable than power density in the free field as it is an indication of the rate of absorption in biological tissues, but the research to set those standards was less than optimum. Truth be told, measurements of electric and magnetic fields are used as surrogates because actual measurement exposures in real life is very difficult due to simultaneously multiple and random EMFS from all directions.

Now that there are transmitters everywhere, a low-intensity continuous exposure is always present—well, almost—especially for those who work near transmitters, the elderly, the confined, and the babies. Does such exposure have biological effects? May it be cumulative? How does one define low-intensity continuous exposure?

Answers to these questions are unknown. One thing is certain though: the guidelines based on SAR do not answer these questions.

If you leave it to the physicists and engineers to tell you if or not there are biological effects of RF, you have set up a design for disaster to begin with. The mannequin of a human skull filled with sugar and salt fluid used to judge whether or not RF would harm living tissue was a grotesque disrespect to science. It was almost a joke. All they wanted to know was how much would heat up 'that stuff' and call one-tenth of that as the safe limit! It is poor science to study if RF exposures will generate an electromotive force by electromagnetic induction using the principle of Faraday's law. That is not how living organisms behave.

Physicists argue that RF in these technologies has such low energy that they cannot influence the equilibrium of chemistry of biological systems. This does not take into account the fact that the chemistry of biological systems is constantly nonlinear and dynamic and not in a state of continuous, perfect equilibrium.

While our understanding about mechanisms that lead to biological effects of EMFs is still a matter for further exploration, the veil has been lifted off the smokescreen created by the industry that technologies based on RF are safe and have no health impacts on living organisms.

Chapter 6

Conundrum of Research

> Deployment of cell phone industry is the largest human biological experiment, ever in the history of mankind
>
> Leif Salford, 2001

Why is it such a big issue? There have been other environmental problems, such as pesticides, insecticides, asbestos, and dioxin. Why the fuss about mobile phones?

Unlike all of the above, the impact of mobile phones is comparable to mammoth issues (like greenhouse effect, global warming, and deforestation) simply because it is practically in the hands of each and every one of us. Besides this is the one that is held closest to the body. What is worse is that it now replaces toys in the hands of infants.

The number of mobile phones may have exceeded human population. In terms of exposure, there is no longer a particular category of at-risk population. We are all at risk—some more, some less—if not from near-field exposure then from far-field exposure. The rapidity with which the technical design of mobile phones, changing bandwidths of infrastructural network, patterns of usage have been evolving in past two decades is yet another issue.

Although there is a whole array of disorders that may potentially result, most people would want to ask, 'Can a mobile phone cause brain cancer?' Short answer, 'Yes, it can.'

Scientific exploration into what happens when there is interaction between radio frequency EMFs and living beings is a highly complex area of research. Radiofrequency microwaves can influence the human body as external electric and magnetic fields couple with the cellular systems of the body. The impact is directly proportional to power density from the source of radio frequency and inversely proportional to distance from the source. Characteristics of RFs—such as frequency, polarization, pulsed versus continuous, distance from antenna, direction of incidence, and the gap between the body and the device—can influence the impact on the body as can certain factors related to the body, such as height, posture, body mass index (BMI), shape of the head, and electric properties of tissues which got exposed (i.e. impact on the bone and cartilage is much less than that on fluids and fat and soft tissues).

Once within the body, the distribution of fields is non-uniform due to differential rates of absorption of different tissues with local hotspots and variations of several orders of magnitude. Rise of temperature up to 0.1 °C can occur in the brain.

Challenges of Research

RF EMF and human health is a contentious subject. By contrast, the task of establishing a link between smoking and human health was kindergarten stuff in scientific parlance. Unlike cigarettes, RF EMF is intangible. Consensus is not lacking on the fact that biological and therefore health effects occur with RF EMF. There is not, however, common ground on how it happens. Even when experimental design and methodology are consistent, interpretation of experimental data may be influenced by the interpreter's beliefs based on conventional wisdom, as has been repeatedly borne out by innumerable studies.

It is not possible to replicate in experimental conditions exactly the real-life exposure to EMFs. In real life, exposures are random, multiple and overlapping with constant variations in frequency and intensity. They also vary from place to place and from time to time. Secondly, naturally occurring reparative mechanisms in the body may also vary with demographic characteristics of the population, depending on a whole lot of variable factors. Thirdly, experimental designs may have inherent flaws. The experimental conditions in the EMF experiments—such as its frequency, impulse shape, duration of exposure, and induced field within the biological target—may affect biological response.

Conflicting biological data might be thus attributable to differences in the frequency and intensity of the field, exposure time, heat generation, cell penetration, and experimental model considered.

Therefore, conclusions drawn from experimental results have to be considered with caution. However, this should not preclude drawing conclusions with some degree of certitude about the impact of EMFs on biological systems.

Results drawn from a particular scientific study may be perfectly accurate within the context of the parameters set out for that study; however, making conclusive opinions about its direct applicability to the human condition may be a risky proposition. It is important to understand how errors can creep in while making such a critical judgement.

Apart from ethical considerations, one of the major difficulties of in vivo research in humans in the field of effects of RF EMF is to compare it with a control group of individuals who have never had a prior exposure. Even if you found people who have never owned or used a cellular phone, their bodies may have been exposed to sources of RF EMF other than personal mobile phones because such radiation is everywhere.

In real-time exposure to RF EMF, the incoming fields are multiple, overlapping, and non-uniform, and their distribution inside the body is not homogenously focused at a point. Instead there are hotspots, and other areas have minimal exposure. It also depends on parameters such as frequency, intensity, polarization, anatomical features of the subject and the type of tissue exposed because not all tissues have similar layers of protection over them. For instance, the eye and the testes have no protection.

In vitro studies pose challenges of their own. For instance, distance between the source of RF EMF, length of exposure, transmission mode (standby versus talk), and presence or otherwise of background RF, the radiation in the lab may influence the outcome.

However, in taking together information and knowledge gained from epidemiological, human, and animal in vivo and in vitro studies, some conclusions can be drawn with certitude. While cellular phones have now become an indispensable part of life, it becomes even more important to be aware of the flip side of their story. To tell people at large that the RF EMFs were safe as long as they didn't cause heating effects in biological tissues was one of the greatest frauds perpetrated by we-all-know-who.

The centre point of research is to identify risks that could lead to disease years later. That is where experimental evidence from animals becomes indispensably relevant. Our bodies get exposed to thousands of cancer-causing agents every single day. It indeed is a miracle if we survive them all, as so often we actually do. It is possible because while the assault on us is continuous, our biological systems have evolved detoxification and reparative mechanisms over the millennia.

Science of Epidemiology

Epidemiology is a discipline of science that studies certain aspects of a disease in a population and contributes greatly to policy decisions in public health. It deals with the distribution of a disease in a population as well as the risk factors such as toxic exposures and their relationship to the prevalence of a disease. Epidemiology employs designs for experimental and observational studies and uses statistical methods to interpret the significance of the results.

An essential component of a classic definition of epidemiology as a discipline is that it measures occurrence, causation, and outcomes of a disease with respect to a sample of exposed *population at risk* in comparison to matched controls. What a wonderful time to be a student of evolutionary biology studying natural selection because the entire world is an open laboratory. The only irony is that it is becoming impossible to eliminate bias from control groups of scientific studies because of the impossibility of finding unexposed controls. Everybody is exposed! That would mean that the studies conducted in the late 1990s may have been somewhat more reliable in terms of quality of control groups than studies in the twenty-first century.

In a cohort study, a cohort is a subunit of a population that shares common characteristics such as one or more of geographic location, ethnicity, age, gender, occupation, or economic status. They are followed over the course of time prospectively against the background of a certain known exposure to determine whether or not they develop the disease.

In a case control study, proven cases are compared with controls while looking back retrospectively at what their exposures were. In a cross-sectional study, exposure and outcome are assessed at the same time, a quick snapshot of the problem.

Evaluation of epidemiological studies is not easy. The goal is to establish a cause-and-effect relationship between the agent and the risk of disease. There are methodological challenges in conduct of research. Errors can creep into studies due to selection bias or even informational (recall) bias, chance, or cofactor confounding.

Unlike in vitro and in vivo studies that look at limited numbers of experimental animals, epidemiology looks at large numbers of real human beings.

Cancer induction due to environment is a very long process. Ideal would be to have known people exposed to an agent and then to be able to follow them up prospectively for long periods of time, 25 or more years, before meaningful results can be obtained and compared with an unexposed group of controls. It is a mammoth task in itself, fraught with several uncertainties.

Another possibility is to look at real patients (cohort studies) who have presumably been exposed and try to establish the link between the exposure and outcome by an analysis of what they recall. This kind of work is very slow and expensive. It's not an easy task either and is fraught with equal uncertainties such as recall bias. Besides, the disease itself may have affected the memory of the interviewee, or the interviewer may be biased one way or the other.

The basic problem with all exposure studies is that collection of data is often imprecise and questionable.

The problem with animal studies is that they are often based on a toxicology model. In case of studies on exposure to RF EMF, how can one quantify exposure, especially when the 'toxin' in question is invisible, multiple, random, and present 24/7? It is not, after all, a cigarette that you smoke! If studies claim that the experiments were done at radiation levels as in human exposure, still the relevance is hard to establish because in humans, it is a matter of lifetime exposure.

Worse still, even those who do not use mobile phones may be exposed to random RFEMFs from so many other sources. Furthermore, the confounding occurs because in real life, people have been exposed to other cocarcinogens through the environment, water, and food.

There is need to rewrite rules of the epidemiological study design. Mechanistic study designs similar to those for toxic exposures of chemicals won't work.

Notwithstanding the limitations of research methodology, an enormous amount of credible work has been done by science academia in last three decades to study the biological effects of radio frequency radiation emitted from wireless technology. Epidemiology has further served to establish the link between bio-effects observed in experimental conditions and health effects in humans, animals, and plants.

Power Lines and Childhood Leukaemia

In the latter half of twentieth century, controversies raged about health impacts of power lines. There is epidemiological evidence of childhood acute lymphoblastic leukaemia and chronic lymphocytic leukaemia in adults from living in the vicinity of high-voltage power lines. Wertheimer and Leeper in 1979 first reported childhood cancer in a population in Colorado living in the vicinity of excessive density of power lines.[1] Milham in 1982 raised similar concerns in people in electrical occupations. Further 200 more studies claimed the same.

Sceptics alleged that no actual measurements of fields were made; there was no laboratory evidence and no known mechanisms of effect on biology. The Energy Policy Act of 1992 in the USA launched the $41 million Electric and Magnetic Fields Research and Public Information Dissemination Program (EMF RAPID), funded partly by the government and partly by the electric utility industry, from 1994 to 1998, which debunked any claims of health risks!

Milham, on the contrary, established a clear link between peak of acute lymphoblastic leukaemia in children aged 2–4 in the 1920s in the UK and in the 1930s and 1940s in the USA in urban environments having residential electrification.[2] Milham went a step further to hypothesize that even cardiovascular diseases, cancer, diabetes, and suicide are not caused by lifestyle but by electrification.[3]

Home use of electricity necessitates that interrupter switches be placed to control the use of electricity. Interruption of the flow of electricity is the cause of the generation of high frequency transients, which Milham calls 'dirty' electricity. Even today, these diseases are virtually absent where there is no electricity. A possible relationship between exposure to EMF and early onset of Alzheimer's disease was also suggested.[4]

Earliest epidemiological studies on health impacts began to appear at the turn of the century. Most of them, expectedly, found no association, yet even then there were some that pointed towards risk of tumour of nerve of hearing and a malignant tumour of the eye.[5]

As early as 2000, Muscat and colleagues ruled out any impact of cellular phones on the incidence of brain tumours.[6] The study was conducted under the controversial $28.5 million WTR (Wireless Telecommunication Research) programme funded by the industry. George Carlo, head of WTR, later commented that the findings of the study were 'fudged' at the behest of the industry and that he had written to the publishers of *JAMA* to retract the study but they went ahead to publish it.

Peter Inskip did not favour the thesis in favour of mobile phone damage either.[7] In 2011 he was known to have walked out of the meeting of scientists in Lyons, held under the auspices of IARC to evaluate the evidence of toxicity of wireless microwave radiation.

In 2004 in a major publication by Siemiatycki et al.[8] about occupational cancer-causing agents, so little attention was paid to health impacts of electromagnetic fields that the authors forgot to include low frequency magnetic fields even though such fields had already been classified as class 2B possible carcinogens by the International Association of Cancer on Research (IARC) under the auspices of World Health Organization two years earlier!

Interphone Study

Prompted by some scientific reports and societal concern, IARC conducted the famous €19.2 million (of which €5.5 million were contributed by the industry), then the largest ever case control study between 2000 and 2004, comprising 5,117 cases of brain tumours—2,708 gliomas and 2,409 meningiomas. It is famously known as the Interphone Study.[9] Its collaboration spanned across 13 countries—namely, Australia, Canada, France, Italy, Germany, UK, Israel, Finland, Denmark, Norway, Sweden, Japan, and New Zealand. One-fourth of the funding came from the mobile phone industry. It would have been most appropriate if the USA was part of the study because most users with the longest period of use of cell phones were in the USA at that time. Purported intention was to find out if the use of wireless RF devices, such as mobile phones, increases the risk of brain tumours and tumours in the acoustic nerve and parotid gland.

The official three-sentence conclusion was this: 'Overall, no increase in risk of glioma or meningioma was observed with use of mobile phones. There were suggestions of an increased risk of glioma at the highest exposure levels, but biases and error prevent a causal interpretation. The possible effects of long-term heavy use of mobile phones require further investigation.'

This ambivalent statement appears to hide more than it reveals. It doesn't deny the risk, yet it tilts towards reassurance that there is no risk. Each to suit his taste. It took six years for IARC to come up with that statement after completion of the study. Despite huge expectations from the Interphone Study, it left most of us still groping for some sense of closure in the smokescreen of doubt. A travesty of science indeed!

Glaring potholes stare at you when you read the Interphone Study report between the lines. For the purpose of inclusion in the study, you could qualify to be called a 'regular cell phone user' if you made a call once a week for a minimum of six months. Cordless phone users were excluded even though in those days, cordless phones were used much more than mobile phones.

Despite the fact that younger people are more vulnerable to suffer health effects, you had be at least 30 years of age to qualify as a participant. At the other extreme, limiting the upper age limit at 59 years would also eliminate the chance of finding glioma in those people who might have been in the latency period. The incubation period for a tumour to develop is usually not less than a decade, yet it was thought prudent to terminate the study in half of that time.

The greatest paradox was that the study reported that regular use of mobile phones might actually protect you from a risk of glioma. The authors published two separate appendices of the study in the same issue of *International Journal of Epidemiology* (for reasons unknown), which unequivocally suggested a 40% increased risk of glioma in the temporal lobe of brain on the same side of the head after at least 1,640 hours of use over a period of 10 years.

Elisabeth Cardis, director of the Interphone Study, was herself a co-author of another study that suggested an association between cell phone use and tumours of the parotid gland.[10]

Subsequent to Interphone Study, a case control study CEFALO was conducted in children and adolescents aged 7–19 years diagnosed with a brain tumour in Switzerland, Denmark, Sweden, and Norway between 2004 and 2008 under the leadership of Maria Feytching. The study concluded that there was no risk of brain tumours even at the highest exposures.[11]

Soderqvist et al. refuted the claim based on flaws in the study design and its statistical interpretation.[12] Notwithstanding that, Feytching has published yet another denial in 2014.[13] In an invitation to make a comment on the effect of extremely low frequency on cancer, Maria Feychting's response was offensively entitled 'Now It Is Enough!'[14]

In scientific studies, it is theoretically possible to determine the outcome beforehand by manipulating the parameters of the study design. 'The percentage of scientific articles retracted because of fraud has increased tenfold since 1975. Retractions exhibit distinctive temporal and geographic patterns that may reveal underlying causes,' says Fang.[15]

This is what happened to the Danish cohort study in 2006, which gave the all-clear to mobile phones and was lapped up by the industry and their trumpet blowers in the media. Up until 1995, the maximal usage of mobile phones was for commercial purposes, and corporate users were carefully excluded from the study. Just as in Interphone

Study, in order to be counted as 'exposed', participants in the Danish cohort study had to make just 26 calls in six months. That is ridiculous by today's standards.

The atomic explosion in Hiroshima in 1945 led to a plume of ionizing radiations, but nobody developed cancer in the first 10 years.[16] Therefore, it is no surprise that studies published at the turn of century didn't show any increase in the incidence.

Finally, Light at the End of the Dark Tunnel!

Lennart Hardell, a Swedish oncologist of Agent Orange[‡] fame, has reported extensively on health risks of mobile phones and cordless phones.[17, 18, 19, 20, 21] He concluded that mobile phones and cordless phones would lead to an increased risk of acoustic neuroma and malignant brain tumour called glioma (especially glioblastoma multiforme grade III–IV).

The risk is highest for those who have used it for ten or more years on the same side of the head. Increased risk of behavioural problems in people living less than half a kilometre from base stations has also been reported.[22]

In their most updated data in 2015, the Lennart cohort study group have conducted pooled analysis of two sets of case control studies on confirmed malignant brain tumours between 1997–2003 and 2007–2009. The patterns of use of mobile and cordless phones of 1,498 (89%) patients prior to the diagnosis were studied. In the case of people who had used mobile phones for longer than 25 years, their odds of developing a brain tumour increased threefold compared to matched controls. Highest odds were found for tumours on the side of the head where the phone is used. The highest risk was for glioma in the temporal lobe. Odds were

---

‡ Agent Orange was an herbicide extensively used during the Vietnam War to destroy dense forests used by counter insurgents as hideouts. Lennart Hardell has the credit for it being recognized by WHO as a possible carcinogen.

much higher when first use of the device started before the age of 20 years.[23] But they failed to find any association with meningioma.[24]

2G GSM mobile phones emit EMFs in tens of mill watts. By comparison, 3G UMTS radiate tens of microwatts that are at least three orders of magnitude lower. Contrary to expectations, however, the risk of glioma is three times higher with 3G UMTS phone.

Residential proximity to mobile phone towers is a documented risk factor for development of cancer 4.15 times more than in the general population.[25]

Though there may be lessons in it for the future, epidemiological studies typically tell the story of the past. Secondly, they reveal at best an association and not necessarily causality. In order to overcome this criticism, Sir Austin Bradford Hill, an English epidemiologist, devised a set of minimum conditions in a presidential address at the British Society of Medicine in 1965 that would support evidence as causality in a scientific study. Now known as Hill's criteria of causation, they are accepted by the scientific community as the litmus test. The criteria include strength, consistency, specificity, temporality, plausibility, coherence, analogy, experimental evidence, and biological gradient. Amidst raging controversy, the criteria were famously the testing ground in case of causal association between smoking and lung cancer.

A common feature of corporate denials throughout history has been to deny consensus and perpetrate a doubt. That was the case with Big Tobacco, pesticides, dioxin, and that is the case with wireless radio frequencies. Lennart Hardell has his own share of critics. In 2001 an article entitled 'Researchers Who Talk Nonsense' appeared in *Svenska Dagbladet*, a Swedish newspaper. The authors, Hans-Olov Adami and Anders Ahlbom, both professors of epidemiology at the Institute of Karolinska in Sweden, made the derogatory reference to Lennart Hardell. Hardell, on his own part, has highlighted the instances of ties between the industry and certain scientists[26] in Sweden, UK, and USA. Nevertheless, his group has set aside the controversy raised by his critics that their earlier studies indicated nothing more than 'association' of brain tumours with use of wireless devices with no causal relationship. Authors have proven that their studies have withstood the test of

Hill criteria and therefore the relationship between RF EMF and brain tumours is indeed causal[27].

It Is Official; EMFs Are Possibly Carcinogenic

Under the auspices of the World Health Organization, the core mission of the International Agency on Research for Cancer (IARC) is to promote cooperation among scientists across the world in research upon the causes and prevention of cancer.

In the last week of May 2011, 31 scientists from 14 countries met at IARC in Lyons, France. They analyzed evidence of studies done up until then and classified radio frequency electromagnetic fields in the frequency range 30 kHz–300 GHz as 'possibly' carcinogenic to humans (Class 2B) based on increased risk of glioma, a form of brain cancer associated with use of wireless phone and acoustic neuroma.[28]

The next morning, the news was headlined all across international media. Of the 31, 1 scientist abstained (unknown), and another, Dr. Peter Inskip (who does not believe that mobile phones cause cancer) of the US National Cancer Institute, walked out.

There was no other dissenting voice. That decision brought it to a new level, said Kurt Straif, organizer of the meeting.

But for the Hardell group of studies, the decision could have been class 3 (risk indeterminate). Anders Ahlbom was initially part of the working group but was removed because of his relationship with his brother's telecom firm. His brother had been a lobbyist for TeliaSonera, a multinational corporate giant in the telecom industry.

The representatives of the telecom industry who were present at Lyons as observers played down the interpretation of IARC classification. Jack Rowley of the GSM Association said, '2B suggests the hazard is possible but not likely.' In its press release, Mobile Manufacturers Forum said, 'It is significant that IARC has concluded that RF EMFs are not a definite nor a probable human carcinogen.' Play of words.

Within four weeks of the IARC publication, ICNIRP published on 1 July 2011 a review entitled 'Mobile Phones, Brain Tumors and the Interphone Study: Where Are We Now?' It shockingly claimed that 'although there remains some uncertainty, the trend in accumulating evidence is increasingly against the hypothesis that mobile phone use can cause brain tumours in adults'.[29]

The IARC classification has given rise to a lot of controversy. It is unfortunate that wireless nonionizing radiation has been placed in the same group as coffee (with respect to cancer of urinary the bladder) and talc-based body powder (perennial use of), both of which have been classified as class 2B carcinogens! It is not surprising that it failed to have any significant impact on policy makers or apply curbs on the industry.

There has been further evidence of association between cell phone use and cancer. Three years after the IARC classification, a CERENAT a multicentre case control study carried out in France between 2004 and 2006 with 447 cases of brain tumours and twice the number of controls

showed evidence of association between brain tumours in people who made heavy use of mobile phones.[30]

Ever since, there have been voices from various quarters, expressing concerns about WHO classification of radio frequency radiation. Scientists have urged IARC and WHO to reclassify radio frequency is class 2A carcinogen. That is where it deserves to be placed because implications of locking it up as 2B carcinogen doesn't make it sound any more dangerous than talcum powder.

In 2013, the Hardell group urged IARC to change the classification. On 30 January 2015, about 5,000 people wrote a petition to WHO/IARC to reclassify RF EMF as class 1. In May 2015, Morgan and his group made a similar appeal.[31]

On 27 May 2016, the National Toxicology Program, under the auspices of the Food and Drug Administration of USA, released partial findings of their $25 million study on the impact on rodents of whole-body exposure to RF EMF, including levels of intensity deemed safe by FCC and used in current cellular telecommunications in USA for a period of nine hours a day starting in utero through their a two-year lifespan. The potential for thermal effects was excluded as no rise in temperature of test animals was found. The control group of animals stayed in exactly identical conditions except that they were not exposed to EMFs and not a single one of them developed a tumour.

The intention was to emulate current exposure patterns in humans, which is what makes this study a wake-up call. At the time of this writing, the final outcome of the study is still awaited, but preliminary results indicate an increased risk of glioma, a cancerous tumour of the brain and schwannoma of the nerves of the heart, especially in male rodents. What makes it further important is that similar tumours have been reported in other studies.

An argument has often been made that if radio frequency microwaves were indeed carcinogenic we would have seen an increased occurrence of brain tumours. Well, the news is that we have. Data from three cancer registries—namely, Los Angeles County, California, Cancer Registry and the SEER 12 cancer registry—suggests an increase in incidence of brain cancer in the regions of the brain where absorption of radiation is maximum—namely, the frontal and temporal regions.[32] In particular, an increase in the incidence of malignant glioblastoma multiforme, a very aggressive brain cancer associated with cell phone use, has been reported from Australia.[33] According to Hans Skovgaard Poulsen of the Danish Cancer Society, cases of malignant glioblastoma multiforme have doubled between 2003 and 2012.[34] He called it a frightening development.

What we have here is a completely new reality. The question of causal relationship between non-thermal effects of EMF and adverse impact on health is foregone. The question that needs to be asked is this: how does it happen at all levels of organization, molecular, cellular, organ, system, and organism as a whole?

Fortunately, current research backs up the explanation of mechanisms of non-thermal impact on biology with firm data and evidence.

Chapter 7

Electrohypersensitivity

In modern medical practice, physicians are increasingly confronted with non-specific symptoms, such as trouble sleeping, chronic tiredness, inability to focus, vague uneasiness, poor energy levels, unexplained aches and pains, lack of control, poor memory, headaches, a feeling of pressure in the ears, ringing in the ears, and anxiety.

The degree of symptoms may be mild but, at times, significantly incapacitating. A clinical diagnosis in the office may be difficult, and exhaustive medical tests may be normal. These symptoms frequently receive a junk diagnosis of chronic fatigue syndrome, neurasthenia, stress syndrome, burnout syndrome, idiopathic environmental syndrome, and sometimes neurosis. Treatments prescribed may include dietary supplements, vitamins, minerals, physical exercise, stress reduction, body massage, and even counselling.

Even in the medical fraternity today, there is not enough awareness about increasing exposure to electrosmog both at home as well as at work, leading to hypersensitivity to radio frequency as one of the likely causes in such cases. One of the reasons is the lack of clearly defined clinical profile or objective laboratory tests to diagnose the condition. Besides, International Classification of Disease (ICD-10) does not recognize it as a disease entity as yet. However, in Sweden, it is recognized as a functional impairment, and it is coded by ICD-10 at W90 as 'exposure to nonionizing radiation'. The international community of scientists aggressively pursues WHO to recognize it is a separate disease entity.

EHS Is Not New

Indeed electrohypersensitivity (EHS) is not a newly discovered entity. As far back as 1932, Erwin Schliephake reported radio wave sickness in the *German Medical Weekly*. Subsequently, in 1970, Zinaida Gordon of Moscow Institute for Industrial Hygiene and Occupational Diseases reported it in her ten-year study with 1,000 personnel stationed at radio installations, such as radar stations and electric utilities. In the Second World War, number of radar technicians suffered from what was called microwave sickness in those days.

The term *EHS* was introduced in 1997 by Bergqvist and Vogel.[1] According to a report by Hallberg and Oberfeld, until 2000, incidence ranged between 0.06% to 3.1% across Europe, but after 2000, it has risen from 5% across Europe to 13.3% in Austria.[2] The extrapolated figures indicate that by 2017 incidence would go up to 50% of the population. According to *Canadian Human Rights Report* in 2007, there were 3% individuals diagnosed with EHS, but with each succeeding year, incidence appears to be rising.

Electrosmog is a colloquial term for a mix of low and high frequency electric and magnetic fields and electromagnetic fields pervading our environment. They arise from a variety of sources such as cordless phones, mobile phones, PC, laptop, iPad, electric home appliances, baby monitors, power lines, and cell towers to name just a few.

All of us, especially in the urban environment, are exposed. Some among us may be more sensitive than others and may go on to develop one or a combination of the following symptoms:

- skin symptoms: itchy skin, a feeling of warmth around the ears while attending a call on mobile or cordless phone
- ear symptoms: increased sensitivity to noise, sense of pressure in ears, ringing in ears, inability to hear clearly Nervous system symptoms: irritability, inability to concentrate, restlessness, poor memory, difficulty falling asleep, anxiety
- cardiovascular symptoms: palpitations, fluctuations in the heart rhythm

- other symptoms: unexplained allergies, loss of libido, dryness in eyes, constant fatigue, aches and pains, etc.

These symptoms are more likely if your usage of mobile phones or cordless phones exceeds average use of more than 30 minutes a day, if you use compact fluorescent lights (CFL) at home or work, if you spend long hours using LCD monitors with wireless connectivity, if you have a wireless router or a mobile phone router in close vicinity, if you use a Bluetooth device in your car, or if you reside near high voltage power lines or substations. They are even more likely if you already suffer from multiple chemical sensitivities (MCS) or heavy metal toxicity.

In mainstream medical parlance, recognition of EHS as a certain clinical entity is still in the stage of evolution. A search for reliable markers that would support the possibility of it being diagnosed as a medical condition is on.

In the meantime, here is what we know.

Three Stages of EHS

One of the earliest complaints in otherwise healthy people with no known illnesses is tingling or burning in and around the ear used to make calls. Most people would then tend to use the other ear. Personnel working in call centres and customer care service centres that use headphones several hours a day are also likely candidates. This may be followed by ringing sensation in the ears, stiffness in the neck, nausea, and itchy sensation on the upper part of the body. If you already have multiple chemical sensitivity (MCS), your risks of developing EHS may be higher. Typically, symptoms tend to disappear on weekends or during vacations when the pattern of use of devices changes. Indeed, at this stage, all symptoms may reverse if exposure to RF is withdrawn.

However, if exposure continues, heartbeat may become audible, pulse may run faster, falling asleep may become a problem, short-term memory may be lost, it may become hard to pay attention where one needs to, headaches may become frequent, urinary urgency may develop, and a constant irritability may draw attention, affecting social relationships. What is worse is that these symptoms may begin to happen with exposure at lower intensities of radio frequency and may tend to persist.

Finally, if exposure continues unhindered for several years, insomnia, chronic fatigue may become established, and symptoms akin to dementia may appear with retrograde loss of memory. If you happen to have diabetes or high blood pressure or autoimmune disorders, their status may be worsened by EHS.

Digital Dementia

The risks are much higher in the case of children. Dr. David O. Carpenter warns against a disaster waiting to happen if schoolchildren continue to be taught with laptops connected to the Internet wirelessly for hours together.

South Korea has one of the highest densities of Internet usage in the world. It has been reported from South Korea that among children, there is increasing incidence of underdevelopment of the right side of the brain, leading to poor development of emotional intelligence and empathy, seriously affecting their ability to build social relationships. The condition has been termed as digital dementia.

Autism has been linked to exposure to radio frequency during pregnancy. This is more likely if a pregnant mother carries a mobile phone in a holster hooked to her skirt or uses a laptop placed on her abdomen. Leaving a baby in a cradle with a baby monitor or an iPad attached to it further increases the risk.

People with known diagnosis of neurodegenerative conditions, such as Parkinson's disease, multiple sclerosis, or epilepsy must be wary of mobile phones.

EHS: A Diagnostic Dilemma

A clear laboratory biomarker for EHS is not yet known. Attempts have been made to look for one. The current hypothesis is that radio frequencies induce hypersensitivity by a combination of factors, inducing inflammation at several places in the body. In one study[3] comprising 1,216 patients of EHS and/or MCS, abnormalities were found in a number of normal parameters of blood and urine that would support the hypothesis that there are physiological differences between healthy and electrosensitive people.

Increased levels of histamine were found in 40% of the cases. Histamine resulting from degranulation of mast cells causes severe itching in the skin and can also increase permeability of blood–brain barrier. Apart from its role in allergy, it is also involved in transmission of nerve signals. Nitrotyrosin, a marker of free radical peroxynitrite, which is known to be involved in oxidative cellular stress, was increased in 28% of the cases. Protein S100B is released largely by astrocytes along blood vessels in the brain, and it was increased in 15% cases, indicating an opened up blood–brain barrier. Evidence of autoimmune process was found in 23% cases by detecting autoantibodies against O-myelin. O-myelin is a protein of the central nervous system. Increased levels of heat shock chaperone proteins HSP 70 and HSP 27 was seen in 33% of the cases, indicating the risk of injury to brain cells by radio frequency.

Metabolites of melatonin were decreased in urine. Evidence of reduced blood flow to the limbic system and thalamus of the brain was found. This may explain why patients of EHS have poor smell, memory loss and poor emotional intelligence. This is the probable cause of digital dementia. Reduced blood flow to the limbic system would ruin growth of personality of a child. Some of the earliest lesions in the brains of patients of Alzheimer's disease occur in the limbic system. This is how EHS might be a precursor of dementia and even Alzheimer's disease.

This clearly shows how neurodegenerative disease may develop by exposure to radio frequency by an interplay of several mechanisms, such as reduction of melatonin, increased oxidative stress, production of inflammatory products by degranulation of mast cells and an easy access to attack the brain through its blood–brain barrier, which has been opened up.

Sunburn by ultraviolet radiation is well known. It induces mast cell degranulation and activation of a protein involved in systemic inflammation known as tumour necrosis factor alpha (TNF a). First discovered by Paul Ehrlich in 1878, mast cell is a type of white blood cell called granulocyte because it contains granules rich in histamine and heparin. Best known for their role in allergies, mast cells also participate in immune defence system, wound healing, and protection of blood–brain barrier.

Radiofrequency EMF also causes changes in the skin similar to ultraviolet radiation through, perhaps, a similar mechanism. Olle Johansson called it screen dermatitis in 1994. He has demonstrated massive degranulation of mast cells in upper layers of the skin in people who sit 40–50 cm from computer screens, leading to itching and redness of the face.[4]

There are several other medical conditions in which these markers may be abnormally altered as well. Although it does not prove that these markers can be used as 'diagnostic' tests, it does lend credence to the fact that EHS is not an imaginary psychiatric illness, which raises concerns to accord respect to these victims of environment. What with more than 7 billion worldwide users of these devices, EHS is fast-becoming a global health problem.

Nikola Tesla (1856–1943), the Serbian American physicist, best known for discovering the AC electricity and induction motor, was probably one of the very few people of his time exposed to high levels of EMF.

John O'Neil, who wrote his biography, mentions weird sets of symptoms that Tesla greatly suffered from during his later years. All his senses became very acutely sensitive to the point of becoming a source of constant suffering. Ordinary sounds became intolerable. A slight touch would send a vibration through his body. His furniture had to be placed on rubber pads to avoid the vibration caused by movement. His heart would beat out of rhythm. He could not withstand strong lights. If he lived today, he would have most likely received a diagnosis of EHS. The other most famous person who suffers from EHS is former director general of WHO, Gro Harlem Brundtland.

Chapter 8

Hormone of Darkness

Sunrise and sunset mark two events in a day–night cycle that govern, through very complex systems, various processes of the lives of plants, animals, and humans. It generally escapes our notice that nature works in a quiet and unobtrusive fashion to facilitate life because it is just so obvious and simple. When the earth rotates around its axis and the sun prepares to go down on western shores of the horizon with the twilight setting in and daylight fading into darkness, a melanopsin-containing[§] set of ganglion cells in the retina of the human eye begin to sense the change in the ambient light and fire an action potential up the tracts of nerves that connect with a specialized group of cells in the foremost part of hypothalamus of the brain named suprachiasmatic nucleus (SCN), which is sitting right atop the point where two nerves of eyesight cross.

This is the first signal to the interior of our bodies that the world outside has turned dark. A sequence of events soon begins to unfold that would affect a lot of processes during the course of the night.

SCN relays a message to a gland shaped like the cone of a pine tree called pineal gland, which prepares to 'cook' a hormone[¶] called *melatonin*, which would be released into blood circulation around two to three hours after the sunset, depending on the time of the year!

Melatonin, the hormone of darkness, lowers your core temperature and blood pressure just a bit, causing your system to relax and gradually

§ Melanopsin is a light-sensitive pigment in light-sensitive cells in the retina of the human eye.

¶ A hormone is a class of signalling molecules produced by endocrine glands, poured into the circulatory systm at precisely defined timings, and carried to distant organs to regulate their function and behaviour.

coaxes you into sleep. The environmental cue, also known as zeitgeber, which is absolutely necessary for this chain of events to swing in, is the onset of darkness. Disturb the cue, and the sequence is disturbed. That is why you feel a sense of disorientation when you travel across multiple time zones in short periods of time as the jet planes of today easily allow us to do. It happens when your body clock is out of synchronization with the time in your destination compared to what it is accustomed to in your original location.

At the arrival of first rays of sun, melatonin machinery begins to wind up and is replaced by the release of yet another hormone, hormone of the day, if you please—namely, cortisol. This raises your core temperature and your blood pressure just a bit as you yawn and stretch out of sleep and 'prepares' you for the day.

The gastric secretions, the thyroxine, and the insulin will soon follow as required. The entire machinery runs with a clockwork precision, especially if you entrain your system with a regular periodicity of the sleep–wake cycle. That is what you call a circadian rhythm.** It maintains a cyclical pattern of physical, mental, and behavioural changes that follow roughly a twenty-four rhythm.

Circadian rhythms allow organisms to anticipate and prepare the system for environmental changes like precision clockwork (e.g. pepsin is released just when it is your lunch time, especially if you have a regular time for lunch. Your mouth begins to water at the smell of food because you need saliva to chew, detoxification enzymes begin to release in the liver just before you eat to detoxify the toxins that will be ingested along with food.

Apart from being diurnal, the rhythms are also periodic—e.g. menstrual and the ovulatory cycle in female species of humans and animals.

So far so good. What happens at the molecular level?

Charles W. Woodworth (1865–1940), American entomologist at University of California, was first to suggest a fruit fly (drosophila) as a model organism for research in genetics because of its similarities with

** The term *circadian* comes from the Latin word *circa*, meaning 'around' or approximate, and *dies*, meaning 'day'.

humans in its circadian rhythms. This was quite a remarkable discovery because it allowed exploration into genetic mechanisms that controlled circadian rhythms. Thanks to this beautiful creature, we have learnt a great deal about how our own rhythms work at cellular level.

In the last two decades scientists have discovered that the circadian rhythm is maintained by a continuously engaged set of genetically driven autoregulatory, reverse feedback systems under the vigil of suprachiasmatic nucleus, which therefore is nicknamed as the central circadian pacemaker or the master clock.

Some organs have somewhat independent additional and parallel rhythms of their own outside the master clock SCN, such as adrenal, oesophagus, lung, liver, pancreas, spleen, thymus, and skin.

Several sources of extrapineal formation of melatonin have been found with the aid of specific melatonin antibodies. These sites include brain, retina, lens, airway, skin, gastrointestinal tract, liver, kidney, thyroid, pancreas, thymus, spleen, immune system cells, carotid body, reproductive tract, and endothelial cells. In most of these tissues, the melatonin-synthesizing enzymes have been identified. Melatonin is present in essentially all biological fluids, including cerebrospinal fluid, saliva, bile, synovial fluid, amniotic fluid, and breast milk.[1]

It was shown by Gene Block that cellular oscillators are both autonomous but also communicate with each other and may collectively produce an electrical output that interphases with the pineal gland to result in release of hormones exactly when needed.

Circadian Rhythm and Biological Clock

Biological clocks and the circadian rhythm are *not* synonymous. Biological clocks are an intricate system of genes and proteins interlocked like cogs and wheels in a reverse feedback loop and that drive the circadian rhythms. The system is essentially driven internally but can be synchronized by external cues, which indeed is a blessing because modern man has to travel. 'Clock genes' contain instructions to upregulate and downregulate proteins in a cyclical fashion. It is usually well synchronized in people who follow a regular pattern in the time they go to bed and wake up. It can get disrupted and lead to disorientation but is flexible in that it can reset itself. Light can turn on and off the genes that control body's internal clocks.

Circadian rhythms can influence sleep–wake cycles, release of hormones (melatonin, cortisol, and insulin), body temperature, and state of mind. They are linked to various sleep disorders, sugar metabolism and therefore diabetes, hunger and therefore obesity and bulimia, bipolar disorder, and seasonal affective disorder.

Apart from a circadian rhythmic drive for sleep and wakefulness, there is also an internal homeostatic drive, albeit smaller, which can induce sleep without inputs from light–dark cycle via SCN and melatonin pathway. This drive will bring on sleep even in the absence of that pathway and appears to 'activate' sleep when you have been awake for long hours. In circumstances when external environmental cues are unavailable, such as a very long flight across multiple time zones, confinement away from sunrise and sunset (such as living in the wrong side of high rise buildings, hospitalization, or imprisonment)—this alternative strategy may be at work to put us to sleep.

Melatonin, core body temperature, and cortisol are used as biomarkers for assessing circadian rhythm. The study of circadian rhythms is called chronobiology. It has brought several insights into our understanding. For instance, timing of medical treatment with body clock may significantly increase efficacy and reduce drug toxicity or adverse reactions. For example, it is wrong to prescribe melatonin in the day and cortisol at night.

Now that the genes controlling biological clocks have been identified in fruit flies, gene upregulation and downregulation and the resultant molecular signals

coordinating circadian rhythms have been studied by changing light–dark periods. This is likely to help scientists improve the disorders referred to above by targeting gene modulations!

They Have It Too

In case of animals and birds, photoperiodism (photoperiod = day length—i.e. physiological reactions of organisms to length of day and or night), which changes as seasons change, is crucial for their survival because they need to have a circadian rhythm in order to be able to predict weather conditions, food availability, predator activity, timing of migration, hibernation, and reproduction. In case of plant behaviours—such as leaf movement, (sunflowers always look at the sun), growth, germination, gas exchange, enzyme activity, photosynthesis, fragrance emission—are driven by their circadian rhythms.

Something Changed That Day and Forever

In twenty-first century, it is hard to imagine that there was a time not too long ago when the only source of light was the sun. On 31 December 1879, in a residential area of Menlo Park, New Jersey, USA, New Year was to be celebrated in a manner unlike ever before. Thomas Edison demonstrated his electric bulb to a few wealthy families, and everyone stood gaping in awe and wonder. He knew though that if his bulb had to reach every house, he needed to 'wire up' every household, and the foundation of electrification was laid.

Something changed that day and *forever!* The rotation of the earth around its own axis since eons allowed life to adapt itself to the rhythmicity of light and dark cycles of day and night. And that wasn't going to be the same ever again. On that day, the boundary that separated day from night became blurred. It was the most challenging time in the life of the hormone of darkness.

Ever since, we have a come a very long way in terms of the applications of electricity. While on one hand, it has brought an immense comfort and security, on the other hand, it has immersed us in an invisible electromagnetic smog that is emerging as the greatest challenge for all forms of living species on the planet.

It was first reported in 1980 that artificial light disrupts secretion of melatonin.[2] It occurs in 99% of people and can shorten the duration of the secretion of melatonin by one and a half hours.[3] Computers, tablets, TVs, and other devices that use LED screens have also been shown to disrupt melatonin.[4] Pictures of earth taken from space show that the intensity of nocturnal darkness is disappearing. The risks are imminent.

Damage to elements of DNA through various toxins—such as naturally occurring terrestrial and cosmic radioactivity, radioactive and non-radioactive toxins that get into our bodies through inhalation or ingestion or through the skin—happens all the time. Biological systems have evolved over a long period of time to develop strategies to detoxify the system and defend itself from such predatory damage. Nature assigned the activity of repairing DNA damage to hours of sleep and developed melatonin for the trick perhaps in order to protect replicating DNA from UV light!

In the last three decades, an emergence of a sea of electromagnetic waves that increasingly surround us at all times of day and night virtually all over the planet poses an unprecedented challenge that the living organisms haven't had time to prepare for. Corrective stance taken by the process of evolution takes millions of years before it cements a self-sustained mechanism of endogenous protection inside the living organism. As our patterns of sleep have become completely delinked from the time of sunset and onset of darkness, the patterns of the release of melatonin in our bodies that nature intends to use as a tool to coax us to sleep have gone completely topsy-turvy. Consequently, our ability to repair damaged cells in our bodies during sleep has diminished.

LED and compact fluorescent lights (CFL) suppress melatonin five times more than sodium lights and incandescent lights. It is the blue wavelength of artificial light in 460 nm[††] spectrum of wavelength that has the worst impact on circadian rhythm. By contrast, the light emanating from fire from burning wood and candle is rich in yellow and red wavelengths. Research on indoor lighting systems is necessary

[††] 1 nm is a billionth of a meter.

to find the kind of lights that would least disrupt melatonin secretion. World Health Organization declared 2015 the International Year of Light and Light-Based Technologies.

The common denominator involved in cellular damage is oxidative stress due to overproduction of free radicals. 'Melatonin is a potent scavenger of free radicals and therefore a powerful antioxidant,' says Professor Russel J. Reiter[5] as well as Poeggeler.[6] It acts as a receptor-independent first-line defence against oxidative damage caused by free radicals and is a broad spectrum antioxidant.[7]

There is enormous evidence that EMFs disrupt melatonin secretion. As a winner of Australia-China Young Scientist award, Dr. Malka N. Halgamuge observed in her study in August 2013 that weak electromagnetic exposure in humans can interrupt circadian rhythm by interfering with melatonin secretion of pineal gland.[8]

An increased risk of leukaemia in people living in close proximity to overhead power lines is well known. It has been suggested that it happens due to disruption in synthesis of melatonin.[9, 10] There is firm recent evidence from a meta-analysis that exposure to magnetic fields carries a strong risk of leukaemia.[11]

The cyclic circadian rhythm of melatonin secretion controls the overall activity of human body from eating to sleep and metabolism.[12] Within the cells, secondary messengers are certain molecules that a cell uses as signals to initiate an activity, such as synthesis of a hormone. One such signalling molecule is *cyclic adenosine monophosphate* (cAMP). Within the cells of the pineal gland, calcium ions regulate cAMP, which then helps convert serotonin to melatonin. EMFs are known to affect Ca^{2+} homeostasis and suppress melatonin activity in a wide wavelength range. Their leakage from cells of pineal gland results in a decrease of the cAMP level and thereby suppression of melatonin.

Melatonin counteracts processes that lead to ageing and neurodegenerative disorders, such as oxidative damage due to free radicals, acute and chronic inflammation, mitochondrial dysfunction, and loss of neural regeneration.[13] In a study of medical students' usage of mobile phones for more than two hours a day, it was found to be linked with decreased melatonin secretion.[14] Reduced secretion of a melatonin metabolite—namely, the 6-hydroxymelatonin sulphate

(6-OHMS)—was reported in workers who were exposed to temporally stable 60 Hz magnetic fields.[15] A sharp drop in melatonin levels in electronic equipment repairers has been reported.[16]

An article that featured on *Global Medical Discovery* (2015 has provided compelling evidence that environmental disruption of circadian rhythms is a risk factor for developing osteoarthritis.[17]

Melatonin is known to protect retina, brain, liver, and sperms against lipid peroxidation. Lack of melatonin exposed sperms to risk of attack by ROS.[18]

Shift Work and Breast Cancer

Provision of 24/7 public services across day and night is a relatively recent trend in history of mankind. Different workers provide services in sets of hours throughout the day and night. Technological infrastructure needs to be monitored twenty-four hours a day. It has led to the concept of shift work. Shift work is fundamentally against the natural physiology as it does not allow circadian rhythms to settle in a pattern. Imagine what happens to the hormone of darkness.

'Nearly 20% of the working population in Europe and North America is engaged in shift work, which is most prevalent in the healthcare, industrial, transportation, communication and hospitality sectors. To date, most studies have focused on breast cancer in nurses and flight attendants. Now more studies are needed to examine this potential risk in other professions and for other cancers,' said Dr. Cogliano, head of IARC Monographs programme.

In a press release on 5 December 2007, the International Agency for Research on Cancer (IARC) announced that 'after a thorough review and discussion of the published scientific evidence, an Expert Working Group convened by the IARC Monographs programme has concluded that shift work that involves circadian disruption is *probably carcinogenic to humans (Category 2A)*'.[19] This is one notch higher than the risk evaluated for nonionizing radio frequency EMFs.

A clear disparity in much higher occurrence of breast cancer in industrialized societies than less developed societies has always been known. Initially, it was put down to change in diet. Now there is increasing evidence to suggest that it may be due to increased exposure to lighting, particularly at night-time.[20]

Breast cancer in women and prostate cancer in men are both hormonally driven cancers, and expectedly increased levels of PSA, a biomarker for prostate cancer, have also been found in men who worked night shifts. Blind women are unlikely to have their melatonin secretion disrupted by artificial light and indeed have been shown to have less than half the risk of breast cancer.[21]

Shift work plays havoc with the rhythm. It is expected that such knowledge will also help find ways on how to keep people at risk of circadian disorders alert during periods of work such as pilots, night-shift workers, emergency room doctors, police personnel, air crew, sailors, and soldiers. Even the design of spacecraft is now being influenced by this knowledge as to how to recreate light–dark cycles in their long journey to help space scientists keep the rhythm.

In airline pilots, rhythms may be disordered. Pilots are legally supposed to have a certain number of hours of sleep between their flight schedules, but those schedules are made according to needs of the airline industry and not according to circadian rhythms.

Chapter 9

Genes Are under Attack

Lying coiled up like spaghetti inside the nucleus of each of hundred trillion of cells in the human body is the six-foot long double-helix DNA, which is the chief component of chromosomes. A gene, the molecular unit of heredity, is a region of DNA that carries a blueprint containing instructions for the protein micromachine in cells called ribosomes to make proteins, the stuff that we are made of. In a manner stranger than science fiction, proteins link up in different combinations and permutations to turn into either liver or lung or heart.

The ribosome itself is made up of 150 proteins, and it takes that many to create one protein. What came first, protein or ribosome, is like the chicken-and-egg mystery. A human chromosome can have half a billion pairs of DNA with thousands of genes. If all of DNA were to be lined up, it would be 30 billion mi. long, which is equal to making a back-and-forth trip from earth to the sun 320 times! To top it all, only 2% of genes code for genes. Remaining 98% has been (unjustly) named 'junk' DNA.

Genes carry the 'code' for creating proteins, the building blocks of our body. This must proceed in an orderly fashion. Imprecision is not an option. DNA damage from environmental factors is repairable to a fair extent and is one of the cellular events taking place all the time but particularly when we are asleep. Alteration in the gene or failure of repair that is sufficient to cause a mutation (a sudden change in the code that will be passed on to next generation of cells during division) can bring about a series of events that can make the entire process of cell growth go awry. Cell proliferation may become dysregulated and apoptosis (naturally programmed cell death) can either become suppressed or exaggerated. These may appear to be small events, but they lie at the heart of genesis of several chronic disorders, including cancer.

Twenty-three pairs of rod-shaped chromosomes, with a member each contributed by the father and the mother, in the nucleus of the fertilized egg contain the code for all the next generations of cells that differentiate to create the entire organism. This includes cells that would eventually create sperms and eggs, thus ensuring that the 'code' will be transmitted to the next generation with uncanny precision. If and when external influences damage this hereditary material in the nucleus of a cell, the next copies of the cell on the assembly line will have garbled information, and derangement in the process will continue to be repeated.

Progeny Is at Risk

The worst place for this to happen is in cells of ovaries or testicles because mutated genetic material in the egg or the sperm will lead to deformity in the offspring. Background radiation, including UV radiation, can cause mutations, but by and large, our bodies have evolved enough to eliminate them by mechanisms of DNA repair.

In the last few decades, the human race has been exposed to several man-made agents with mutagenic properties, such as chemicals, pesticides, insecticides, cigarette smoking, asbestos, certain pharmaceutical drugs, and ionizing radiations. Modern man is constantly up against a multitude of mutations in the midst of ever-increasing technological advances. Regardless of what led to it, gene mutation is the beginning of neoplastic change.

Of all the mutagenic technological advances, one that is more pervasive than any other and one that affects every single living entity—man animal, or plant in the entire planet—is man-made nonionizing RF EMFs in the shape of wireless communication systems along with the infrastructure laid out to support it. It is a yet another 'layer' added to our atmosphere, covering the entire planet. Were a particular drug or food found to be toxic to the genome, it would not have been allowed to be marketed.

Age-old wisdom has that children born to younger parents are likely to be healthier than those born to older parents. Spermatocytes and eggs are present in the male and the female child respectively at birth, and they mature to produce fertile sperms and eggs at puberty. If spermatocytes mutate prior to puberty, the mutation will continue in further generations of cells and lead to either male infertility or congenital disease in the progeny. Male children are more likely candidates to suffer than their female counterparts.

In the case of effects of ionizing radiations on biological systems, conclusion as to whether and how these radiations produce biological effects was easy and straightforward. They had enough energy to cleave through chemical bonds by pulling electrons away from molecules that make up tissues. The nuclear explosions in Hiroshima and Nagasaki in 1945 continue to stand a testimony to that even today. It is not as straightforward in the case of nonionizing radiations as the energy associated with them doesn't do so quite in the same way or with a similar force. But at the same time, there is overwhelming evidence that there are effects, or else the precautionary principle wouldn't exist! The effects, in this case, are indirect and secondary to the induction of biochemical changes in cells.

Such was the enthusiasm about X-rays when they were newly applied in medicine that before discovery of anti-tubercular drugs, one way of treating the lung affected with tuberculosis was to collapse it intentionally and to later reinflate it after a period of time. This was achieved with the help of X-rays, and many times, this could only be achieved by repeating it several times. In this group of patients, incidence of breast cancer on the same side as the collapsed lung was found to be four to eight times more than the rate otherwise.

Science has developed tools to eavesdrop on what the cells are doing inside them and techniques to interpret what those tools are telling us. Molecular biology has taught us that the fundamental change that initiates cancer is mutation in the gene, resulting directly due to the effect of a toxin and failure of DNA to repair itself. Such evidence is reliable because it is replicable.

Literature is replete with cohort, case control, animal in vitro, animal in vivo, cell line, and human epidemiologic studies that all point to one direction: RF EMF is not just an incidental association but a causal agent leading to tumours in the brain. Long before mobile phones were heard of, in 1970s and 1980s, workers exposed to microwave radiations would often complain of headaches, dizziness, memory loss, and fatigue.

US Air Force funded a $4.5 million study led by Arthur William Guy, popularly known as Bill Guy, on 100 male rats which were exposed to low-level microwave radiation (0.48 mW/cm^2 at 2,450 MHz; 8 Hz modulated, pulsed microwaves, 10 μs pulses for 21 hours a day for up to 25 months). The maximum average SAR that the rats were exposed to was 0.4 W/kg. The conclusion—microwaves *promote* cancer!

He was an advocate of the use of SAR to measure RF exposure. Bill Guy died on 20 April 2014.

EMFs Damage DNA

DNA damage as well as repair is a normal consequence of being alive. If errors outnumber capability of self-repair, mutations accumulate to a point that they can suppress DNA repair genes (such as tumour suppressor genes) and activate oncogenes that promote cancer. The commonest form of injury to DNA is strand breaks, which may be single-strand breaks or double-strand breaks and DNA cross links. Single-strand breaks are more common and easier to repair, whereas double breaks are usually lethal for the cell. Comet assay is the technique to detect such breaks.

Some of the earliest insights into what RF EMF could do to human tissues came from the foundational work of Lai and Singh from the University of Washington, Seattle, USA, in 1995. The study design was to look for evidence of DNA breaks via alkaline micro-gel electrophoresis after exposing cells derived from rat brains in an in vitro setup. In the first phase, they exposed the cells at 2,450 MHz, using a pulsed signal (500 pulses/sec each pulse of 2 μs width), for two hours, but nothing happened. But when they changed dose rate such that the cells received exposure at 0.6 and 1.2 W/kg whole-body SAR for four hours, single-strand DNA breaks were detected. When exposed to continuous signal at 2,450 MHz, breaks were observed both immediately and at 4 hours after exposure.[1]

Further they showed that there was no significant difference in the outcome with pulsed or continuous stimulus.[2]

Lai and Singh also demonstrated that acute two-hour exposure to 60 Hz ELF EMFs at 0.1, 0.25, and 0.5 mT could also lead to DNA strand breaks. While 0.1 and 0.25 mT caused single-strand breaks, 0.25 and 0.5 mT caused double-strand breaks[3].

In subsequent experiments, they demonstrated that these effects could be prevented by prior administration of antioxidants, such as melatonin and N-tert-butyl-alpha-phenylnitrone (PBN), which opened the door to a possibility that free radicals could be responsible in the causation of RF EMF effects.[4, 5]

They also studied the effects of the magnetic field. They exposed rats to 60 Hz magnetic fields at 0.01 mT for 24 hours and found significant increase in both single-strand as well as double-strand breaks. If exposure was extended to 48 hours, there was greater damage, suggesting cumulative effect. When rats were protected with antioxidants (trolox, a vitamin E analogue, or deferiprone, an iron chelating agent) such effects could be prevented, suggesting a role of free radicals in causation. Acute magnetic fields increased apoptosis and necrosis in brain cells.[6]

It is well known that DNA damage underlies several neurodegenerative disorders and therefore these studies are very significant.[7]

A number of other laboratories failed to reproduce similar results. One of them was at Washington University, led by Roti Roti. Curiously, Motorola funded this study.[8] Yet another study has dismissed the notion that wireless RF radiation can cause genetic damage.[9, 10]

Differences in study outcomes often arise from differences in conditions of the experiment. The effect depends on the absorption of energy by the biological system, which in turn depends on temporo-spatial conditions of exposure—namely, frequency, intensity, duration, modulation, of signal. The outcome also may vary according to the cell type used as different cell types and organisms may respond differently. Technique of detection is also important. Comet assay has several versions, and all of them do not have the same sensitivity.

After Lai and Singh showed the way, there have been several researchers who have shown evidence of DNA breaks as a consequence of exposure to RF EMF in in vitro experiments in various types of human and other mammalian cells.

Some examples are as below:

- Diem et al. showed single and double breaks in human fibroblast and rat granulosa cells.[11]
- Gandhi and Anita showed DNA breaks and micronuclei in lymphocytes obtained from cell phone users.[12]
- Nikolova showed damage in mouse embryonic stem cells.[13].
- Lixia reported damage in human eye lens epithelial cells.[14]
- Aitken showed it in the spermatozoa of rats.[15]

Although RF EMF emitted from cellular phones have received much attention, it is important not to lose sight of the fact that these devices also emit extremely low frequency magnetic fields.[16] This is even more important because newer generations of LTE-class of phones have more powerful batteries creating magnetic fields. In 1997 Linde reported magnetic flux density of 1.8 μT with digital GSM phones.[17]

Azadeh Hekmat reported that the radio frequency from 940 MHz cellular phone at 40 mW/kg SAR could lead to DNA disaggregation in calf thymus DNA immediately after exposure.[18]

REFLEX Study

REFLEX study was conducted between 2000 and 2004. Twelve research groups from seven countries collaborated under the leadership of Franz Adlkofer of VERUM Foundation, Germany. It was designed to study if genotoxic effects of ELFs and RF EMFs described earlier by Henry Lai and N. P. Singh and others could be replicated with biological material from 'human donors' in an in vitro setting in the laboratory by exposing cultured cell lines to ELFs and RF EMFs.

A word about cell lines: It is a particular kind of cell derived from a biopsy; it can be cloned over and over again to produce virtually an assembly line of identical cells called cell lines. They have advantages over primary human cells for experimental purposes in that they provide a genetically homogenous background that is crucial for studies and that they can be stored indefinitely in liquid nitrogen and therefore are a source of abundant supply for genetic material (DNA, RNA, proteins). An added advantage is that with studies on them, there is a better chance of replicating results by eliminating variability.

Without a shadow of doubt, REFLEX study concluded that RF EMF can modify the expression of several genes and proteins and are genotoxic. Effects on cell proliferation and differentiation and apoptosis are less clear. Results are from in vitro studies on human cell lines and reflect impact at the molecular level. Whether or not it confirms a health risk from EMF exposure is still an open question, but in vitro results would rather suggest that the risks are real.[19,20] Human fibroblasts were exposed to extremely low electromagnetic fields (ELFs), and DNA strand breaks were observed at intensities as low as 20 μT after 20 hours, and when intensity was raised to 35 μT, such effects became obvious at 15 hours. Chromosomal abnormalities such as gaps, breaks, rings and dicentric chromosomes were observed. In case of non-thermal microwave RFs, single- and double-strand breaks and worse chromosomal abnormalities were seen at a SAR as low as 0.3 W/kg in human fibroblasts, HL-60 cells, and granulosa cells of rats.

Under the auspices of REFLEX study in an in vitro setting, Hugo Rudiger and colleagues studied the effect of exposure of cultured fibroblasts and lymphocytes from human donors to 1,950 MHz UMTS (Universal Mobile Telecommunication System) radio frequency below 2 W/kg. They looked at evidence for effect on genes by a comet assay and micronucleus assay under blind control. They observed increased comet tails as well as micronuclei after 24 hours at 0.05 W/kg SAR and after eight hours when they increased SAR to 0.1 W/kg in fibroblasts. This is despite the fact that fibroblasts are supposedly least sensitive to non-thermal microwaves. No effect was observed on lymphocytes.[21]

In his comments on this study, Alexander Lerchl raised doubts about the accuracy of statistical calculations. Rudiger rebutted the claim. The spat between the two has been alluded to elsewhere.

Franzellitti, Guler, Kesari, and Karaca have successfully replicated similar effects.[22, 23, 24, 25]

Proteins are building blocks of the human body. Their fundamental structural units are amino acids. The secret of their becoming functionally active in diverse ways and performing diverse actions lies in the ability of amino acids to join up in an innumerable array of designs and shapes. This is called protein folding. Now there is evidence that RF EMF can non-thermally affect this process in nature, resulting in potential for disaster. What if it happened to enzymes involved in DNA repair?[26]

In another in vitro experiment, authors exposed germ cell lines derived from mouse spermatocyte to RF EMF to study genotoxicity. RF parameters were 1,800 MHz GSM signals 24 hour intermittent exposure with 5 minutes on and 5 minutes off at SARs of 1, 2, 4 W/kg to simulate real-life exposure. There was a definite increase in the reactive oxygen species, and these phenomena could be mitigated by pretreatment with alpha tocopherol, an antioxidant. They did not find DNA strand breaks, but they did find evidence of damage to DNA base.[27]

Micronuclei

Abnormalities of the nuclei of cells are a very important indicator of current or impending disease. Micronuclei (also known as Howell-Jolly bodies) are tiny imitation of nucleus seen in a newly formed cell; having a chromosome of its own, either in part or whole, it is therefore capable of creating abnormal copies in the future. It is a certain evidence of damage to DNA that has failed to repair itself. It can result from exposure to environmental factors, such as chemicals and radiation. It is typically found in cancerous cells. An in vitro Italian study showed that non-thermal microwaves at low intensities lead to the formation of micronuclei in human lymphocytes.[28]

Tissues in our bodies are continuously renewed throughout life. Stem cells are the source for maintaining this normal turnover to keep regenerating organs, such as skin, blood, stomach, and intestines. They are themselves undifferentiated and keep renewing by cell division, maintaining their undifferentiated status, but when called upon, they can differentiate into the type of cells of the organ where they are located.

Stem cells are the most sensitive to non-thermal microwaves. A study showed that exposure to GSM/UMTS microwaves inhibited the formation of endogenous 53 BP1, a binding protein in mesenchymal stem cells derived from fat cells.[29, 30] The 53 BP1 has a crucial role in end-joining the double-strand broken DNA. Its inhibition severely diminishes the capability of broken DNA to repair itself.[31]

This leads us to yet another question.

What causes damage to the DNA?

That takes us into the quantum world of organic biochemistry at the level of subatomic particles, where myriads of activities are taking place continuously to keep us alive. Biology depends on chemistry, and chemistry depends on physics. Let us then delve a little into the quantum world of how movements of electrons affect changes in cellular behaviour.

Chapter 10

Radiofrequencies Cause Cellular Stress

A better-known physiological mechanism of what we call stress is based on neurological and hormonal events that occur throughout the entire body via the hypothalamus-pituitary-adrenal axis and manifest as the increase in respiration, heartbeat, and muscular energy, pumped by release of adrenaline and cortisol.

There is another kind of very subtle stress that takes place at cellular level without any immediately perceivable effects but is one that may lay the foundations for a chronic disease process, including cancer. It is oxidative stress induced by free radicals.

In case of ionizing radiation, such as from X-rays and CT scans, the answers were quickly clear but not before human casualties occurred, such as in the case of radium dial painters. Ionizing radiations have enough energy in them to break chemical bonds between atoms and molecules. In the case of RF EMF, historically scientific pursuit was similarly modelled to learn if energy transferred by RF EMF into biological tissues exceeded random thermal molecular collisions. In retrospect, it appears that this Descartean reductionist approach was too naive. Biological organisms are far too complex for that, and as new paradigms in physics have revealed, quantum phenomena at cellular level are far too subtle.

Father of Free Radical Theory

In the history of science, 25 November 2014 was a sad day; it was when Tom O'Connor, spokesman for the University of Nebraska Medical

Centre, announced, 'Harman (Denham), a six-time nominee for Nobel Prize, professor of biochemistry and internal medicine at University of Nebraska Medical Centre (UNMC) in Omaha for over half a century, who worked even in his mid 1990s at the university, arriving each day at 7 a.m. has died after a brief illness.'

Denham Harman was the father of theory of free radicals. His brief but landmark paper in 1956, 'Aging: a Theory Based on Free Radical and Radiation Chemistry',[1] has been an inspiration for a generation of scientists working on the subject of ageing. He had been working at Shell Oil on the chemical properties of free radicals. At the time, it was not known if free radicals existed at all in the human body.

Free Radicals

An atom is the smallest unit of matter, composed of nucleus and electrons. Of late, though, physicists have discovered an array of several subatomic particles. A nucleus is composed of positively charged protons and electrically neutral neutrons held together by strong nuclear force. Protons and neutrons are components of nucleus, and together are called nucleons. Over 99.94% mass of an atom is nucleus. The nucleus is surrounded by orbits of negatively charged electrons spinning in pairs in opposite directions. For an atom to be electrically stable, the number of protons and electrons should be equal.

An atom is called an ion if it has either deficit or surplus of electrons. If the number of electrons is less than the protons, it is positively charged *cation*. If the number of electrons is more than the protons, it is negatively charged *anion*. The electromagnetic force is the attraction between the electrons and protons of its nucleus.

Electrons are the substance that binds atoms together to form molecules. All electrons are charged, and hence, they attract ions with opposite charge and repel those with similar charge. When moving, they follow trajectories which can be deflected by a magnetic field. The electron cloud of an anion is larger than the parent atom. The electron cloud of a cation is smaller than the parent atom.

The affinity of an atom to bond with the neighbouring atom is called its valence. It is a property conferred upon an atom by the configuration of the electrons in its orbitals, which may contain zero, one, or more electrons. For an atom to be stable, it is required that its electrons should equal in number to its protons in its nucleus and that electrons should be in antiparallel spinning pairs. In this state, the atoms would be stable and would have the least energy and reactivity. Electrons spin in pairs, and so long as each electron has its partner, the atom is said to be in 'ground' state.

Atoms that make up all materials, including the human body, are programmed to maintain a neutral charge by keeping the number of electrons in them equal to protons. Electrons are the currency of all materials that keep getting exchanged to regain stability. Most of us have experienced the phenomenon of static electricity. When any two materials come into contact, a stream of electrons may be on the move! If you sit in a chair or lie under a blanket and your body has picked up electrons from that surface, you are carrying a net negative charge. Now if you touch a doorknob and if that knob has positive charge due to inadequate electrons, some electrons will jump from your hand to the doorknob, and you will feel what is called static shock! If you rub a comb on your hair and then hold that comb next to a water stream, the water stream will bend in an attempt to steal electrons from the comb. This is more likely to happen in cold weather with dry air. In summer, the humidity in the air helps electrons move away, and you won't pick up static charge so easily.

The chemical behaviour of an atom depends on one particular structural attribute (i.e. the number of electrons in its outer shell). If an electron becomes unpaired (due to loss of its partner), the atom becomes unstable. An unstable atom will frantically try to find an electron from somewhere in order to regain stability and is therefore highly reactive. This is called free radical. It would naturally try to strip an electron away from a neighbouring molecule, causing that molecule to become a free radical, and the chain goes on.

Understanding the concept of free radicals may be misleading. Most definitions given in literature would lead one to believe that a free radical is an exceedingly tiny, highly reactive and toxic, oxygen-centred molecule that will be less toxic if one electron was added to it (or removed from it). While most of it is true, it is not necessarily exceedingly small. Some are made of only two atoms, while some are huge, resulting, for example, from removal of one electron from a protein or a chromosome.

A molecule is a free radical if it possesses unpaired electrons in its orbitals. Pairing, by the way, implies a set of two electrons spinning in opposite directions. Therefore, being unpaired does not necessarily mean an odd number of electrons. Thus, a molecule can have two unpaired electrons as well, spinning in the same direction. Such a molecule is called diradical—e.g. molecular oxygen (O_2). In this state, oxygen is not reactive and is said to be in ground or triplet state. It can be excited into singlet state by lighting a match or by irradiating oxygen with X-rays. Free radicals centred on oxygen molecules contain two unpaired electrons in the outer shell of the atom and are called reactive oxygen species (ROS), such as superoxide, hydroxyl, hydrogen peroxide, and singlet oxygen.

The formation of free radicals is a two-edged sword. Some systems in the body produce free radicals as part of their own normal activity. For instance, the immune system produces them and uses them to mark invading bacteria and viruses by stealing their electrons for subsequent deactivation. However, if that sort of assault happens on a person's own normal tissues, a disorder in that tissue is bound to set in. This is one way in which autoimmunity begins. The concept of using antioxidants to neutralize free radicals rests on the same theory, whereby antioxidants happily donate free electrons to roaming thieves (free radicals) that could otherwise be harmful. Antioxidants can donate an electron without becoming unstable themselves.

It is, however, questionable if antioxidants administered from the outside would also interfere with molecular mechanisms wherein formation of free radicals is just a step in a beneficial chain of events. Besides, the body has its own endogenous antioxidants, such as melatonin. The body can manufacture antioxidants when we eat fruits, vegetables, nuts and seeds, meat, and oil. Under physiologic conditions, if the body's antioxidant system is jeopardized, free radicals may assume a position of potential harm, leading to oxidative stress. This could lead to tissue injury. This is why a period of rest is necessary between sessions of heavy exercise, during which the excessive production of free radicals occurs due to massive oxygen consumption.

Denham Herman (1916–2014) first proposed in 1956 that a species of highly reactive molecules which are freed in the midst of normal chemical processes in the body unless deactivated by antioxidants (natural or otherwise) cause disease and ageing by their destructive actions in cells and tissues.[1] For several years, his theory was ridiculed and dismissed, but in the 1960s, it gained support. The rise in the incidence of cancer and cardiovascular diseases in the twentieth century was attributed to oxidative stress induced by free radicals. Dr. Herman's further work in elucidating the role of antioxidants (such as Vitamin C, E, and beta-carotene) has remained a landmark in the history of science.

According to free radical theory of ageing, getting old is the result of the overaccumulation of unused free radicals over time, which hasn't been neutralized. Free radicals are everywhere—in the air, our bodies, and the materials around us. They cause the deterioration of plastics, the fading of paint, the degradation of works of art, buildings, cars, and ageing-related illnesses, and they can contribute to heart attacks, stroke, and cancers. In our bodies, we produce more free radicals when we are stressed, eat fatty foods, smoke, and drink alcohol.

In living beings, chemical reactions occur by means of transfer of electron. When an atom, ion, or molecule loses an electron, it is said to be oxidized, and when it gains one, it is said to be reduced. This reduction and oxidation is termed as redox reaction.

Oxygen is available everywhere inside the body, and it readily accepts an electron, thus converting into a free radical, and this is why an opportunity for the formation of free radicals is immense. Free radicals are frantically in the hunt for a free electron and get attracted to anything in their vicinity with an agility of iron filings getting attracted to a magnet. If the collision occurs with proteins and lipids in a cell membrane or DNA, the damage they leave in their trail is serious. If the collision occurs with an invader, such as bacteria or virus or a mutated and potentially cancerous cell, the damage that follows may actually be beneficial.

Reactive Species

The two major types of radicals in humans are reactive oxygen species (ROSs) and reactive nitrogen species (RNSs). The generation of ROS and RNS is a normal phenomenon in the body for useful purposes. The majority of reactive oxygen species are produced by aerobic metabolic processes. They contribute to immunological defences by attacking various pathogens and are also involved in transduction signalling pathways involving voltage-gated calcium channels.

The damage caused by them is continuously repaired. As long as balance is maintained between their production and neutralization, no harm is done. If this balance is disturbed, as in the case of EMF exposure, damage and disease can occur. Natural antioxidants keep their level in check. If they did not, free radicals would attack lipids, nucleic acids, and proteins, resulting in oxidative damage. There is evidence that EMFs can amplify free radical generation multifold. ROS are more prevalent than RNS and are formed by iron-mediated Fenton reaction (Heber–Weiss Fenton reaction to be exact) by producing the hydroxyl radical, which is the spookiest of them all. This can lead to DNA breaks.

An upshot of lipid peroxidation caused by free radical cell membrane damage (regardless of how it started) is the release of melanodialdehyde (MDA) and 4-hydroxynonenol. These can cause havoc by inactivating membrane enzymes, disturbing protein–lipid interactions in membranes, forming intermolecular cross links, changing the viscosity of the lipid fraction, and preventing the formation of enzyme substrate complex.

M. Simko et al.[2] in a hypothesis suggested that EMFs induce an activated state of the cell (e.g. phagocytosis, signal transduction pathways via Ca^{++} channels), which then amplifies the release of free radicals, leading to genotoxic and other cellular damage.

Taken together, this would explain DNA and other cellular damage. Free radicals affect cells by damaging macromolecules, such as DNA, protein, and membrane lipids. Several reports have indicated that EMF enhances free radical activity in cells particularly via the Fenton reaction. The Fenton reaction is a process catalyzed by iron, where hydrogen peroxide, a product of oxidative respiration in the mitochondria, is converted into hydroxyl free radicals, which are very potent and toxic molecules.

The energy content of ELF EMF is many a magnitude lower than that of RF RMF. Interestingly, they too affect DNA, and there is evidence that they do so via free radicals.

The generation of free radicals depends on how much hydrogen peroxide is produced by the mitochondria to fuel the reaction. Therefore, cells that are metabolically more active would do so much more and are therefore more susceptible to EMF.

Cells with higher levels of intracellular iron would be more vulnerable to EMF. Brain cells have high levels of iron. Rapidly multiplying cells, such as those in a cancerous tumour, have higher uptake of iron. While the whole organism is exposed, EMF might selectively affect these cells more than others. There are two implications: Can EMF be used to kill cancer cells? Is that why, in an Interphone study, one report said that EMF actually protected against risk of glioma?

Likewise, DNA breaks are more likely to appear in the brain than elsewhere. Brain cells have lesser capacity to mount DNA repair. Hence, neuronal damage is a bigger risk. As nerve cells do not divide, cumulative DNA damage is likely to lead to neurodegenerative disease, whereas glial DNA damage can lead to cancerous disease.

Aneuploidy is a major somatic mutation in which the number of chromosomes in daughter cells during cell division turns out to be abnormal, either less or more than the expected 46. It is known to lead to instability of genome and therefore disease. RF EMF at subthermal levels has been shown to cause aneuploidy in cells.

Oxidative stress has been implicated in the pathogenesis of multistage carcinogenesis, autoimmune diseases, cardiovascular diseases, neurodegenerative diseases (Parkinson's disease, Alzheimer disease, Lou Gehrig's disease, Huntington's disease, and cerebral ischaemia), mitochondrial and respiratory diseases, infertility, Down's syndrome, autism, ulcerative colitis, rheumatic arthritis, irritable bowel disease, atherosclerosis, and ageing. ELFs have been linked with childhood leukaemia.

So far, evidence of free radical activity is through the measurement of their biological effects—namely, lipid peroxidation, DNA damage, protein damage, perturbation of enzymatic/non-enzymatic antioxidant defence—and then on the reversal of all these effects by natural and artificial antioxidants (melatonin, ROS spin traps, etc.). This evidence is indirect.

Naturally occurring antioxidants include catalase, superoxide dismutase, and glutathione apart from melatonin. Direct biomarkers of oxidative stress include increased levels of melondialdehyde (indicative of lipid peroxidation) and nitric oxide and reduced levels of glutathione (GSH), superoxide dismutase (SOD), catalase GSH-Px, and pro-oxidant enzyme xanthine oxidase (XO).

The Calcium Ion

Survival of a living organism necessitates that cellular systems of the entire organism communicate with each other. For instance, in the face of danger, when you need to run, you cannot do so without your brain, heart, lungs, and muscles participating in the feat at once! Even at rest, maintaining a continuity of homeostasis is fundamental to development, tissue repair, and immunity. Communication between cells requires a mechanism for signalling through messengers. Molecules that act as signalling messengers include hormones, neurotransmitters, cytokines, growth factors, chemokines, protein kinases, and components in extracellular matrix.

At the molecular level, nature has adopted positively charged Ca^{2+} ions and negatively charged phosphate ions as two fundamental signalling elements that bind to hundreds of proteins to trigger intracellular changes in gene and protein expression that affect every aspect of life and death of a cell. In addition to being intracellular messengers, calcium ions sit between phospholipids that make up the cell membrane to provide adhesiveness. Movement of voltage-gated Ca^{2+} ions from the extracellular compartment to the intracellular compartment changes membrane potential.[3] Errors in this complex process may trigger far-reaching consequences, resulting in autoimmune disorders and cancer.

Over two decades ago, J. Walleczek[4] suggested that weak EMF-altered Ca^{2+} regulation is an early trigger of field effects in the cells of the immune system. Weak ELF magnetic fields can modify Ca^{2+} regulation in lymphoid cells, which is of paramount importance in normal proliferation of lymphocytes, possibly mediated by EMF interference with Calcium-dependent signal transduction mechanisms on lymphocytic proliferation. This is a remarkable discovery of immunology! This has nothing to do with thermal effects. A similar responsiveness to weak EMFs has also been documented for cells and tissues of neuroendocrine and musculoskeletal systems, indicating that EMF sensitivity might be a general property of biological systems. That was in 1992. Considerable progress has been made ever since in further elucidating mechanisms involved.

Ninety-three percent of studies up until 2015 indicate that low intensity RF EMF cause oxidative damage mediated through free radicals.[5] That says a lot about the bioactivity of free radicals in mediating cancer and several neurodegenerative disorders.

In 2013 Martin Pall[6] has shown that nonionizing EMFs—including ELF EMFs, MW and RF fields, nanosecond EMF pulses and even static and magnetic fields—target voltage-gated calcium channels (VGCCs) by altering membrane potential, which causes VGCCs to massively increase production of intracellular calcium.[7, 8] It then stimulates enzyme nitrous oxide synthatase that produces the reactive nitrogen species, namely nitric oxide (NO). One of two things may happen from here on. An NO/cGMP (cyclic guanosine monophosphate) protein kinase pathway[9] may be activated, leading to a beneficial stimulation of osteoblasts and leading to bone growth (such as in healing of a fracture), or an NO/peroxynitrite pathway may be triggered, in which the NO reacts with superoxide to form peroxynitrite, which can produce free radicals, such as hydroxyl (HO) and NO_2 radical.[10] Polyunsaturated fatty acids in the membranes of cells are submissive substrates for these free radical species to attack, leading to an explosion of cellular integrity. Consequences are far-reaching. It is of interest to note that calcium channel blockers, such as verapamil, can block such effects of EMF.[11]

The oxidative stress resulting from peroxynitrite has been shown to cause single-strand DNA breaks,[12] and the fact that some antioxidants may have a protective role against this lends further support to this finding.[13] The induction of heat shock protein (HSP 70) in response to ELF magnetic fields as low as 0.025–0.10 mT has been shown to be inhibited by the free radical scavenger melatonin and clearly demonstrates the involvement of free radicals as an oxidative stress response to these fields.[14]

Calcium regulation plays a significant role in the proliferation of lymphocytes, one of the key components of cellular mechanisms of the immune system. In the 1990s, experimental evidence had shown that calcium regulation gets altered as a result of EMF exposure.[15, 16]

In 1996, in a landmark article published in *International Journal of Radiation Biology*, Henry Lai and N. P. Singh showed that single-strand as well as double-strand breaks in the rat brain cells occurred after an acute exposure of only two hours with continuous as well as pulsed electromagnetic waves.[17] In 2004, in yet another important study, they demonstrated that if the rats were pretreated with antioxidants such as melatonin or trolox (vitamin E analogue) prior to exposure, the damaging effects on brain cell DNA could be blocked![18]

This opened up a whole field for study as it suggested that the free radicals had a role in mediating damage to brain cell DNA.

Oxidative Stress Does It All

In a further article published in September 2004, Richard G. Stevens,[19] while applauding Henry Lai's work, mentioned, 'Free radical burden and oxidative stress therefrom may be a unifying theme of adverse effects of EMF on all biological consequences.' He postulated that EMF-induced loss of iron from ferritin (the intracellular protein, which is a stored version of iron) is one and that the second is increase in free radicals due to EMF-induced changes in calcium homeostasis.

As our body contains a large amount of water, it is reasonable to expect that its interaction with EMFs will have a bearing on biological effects. The proposed mechanisms of interaction between the ELF EMFs and the biological tissues include direct acceleration of ions in biomolecules and including those of DNA and distortion of gating of electro-sensitive channels on cell membrane.[20]

The interaction in the case of ELF must be via the electric or magnetic component. In the case of ELF, polar organic molecules of human tissue attenuate the electric component, but the magnetic component, however small, remains unchanged and is the likely source of interaction. Indeed, this has been supported by epidemiological studies with magnetic fields stronger than 0.4 mT superimposed over geomagnetic fields[21] and direct biochemical evidence with 100 mT on murine fibroblast derived periadipocytes and rat brain cells via free radicals.[22] The latter occurs through free radical mechanism. Free radicals have a very short lifespan because they are so highly reactive that they would react with biomolecules instantly and change their chemistry and behaviour.[23]

There is evidence that extremely low frequency EMFs generated by power lines are also capable of causing DNA breaks via generation of free radicals. These effects can be blocked by prior administration of antioxidants, such as melatonin.[24, 25]

In 2002, IARC has classified ELF magnetic fields as class 2B (possible) carcinogens. Paradoxical as it may seem, in the same year, IARC classified ELF electric fields as class 3 (indeterminate), whereas we know that the former cannot occur without the latter! RF has similar effects. In two studies by Lai and Singh, in rats exposed to 2,450 MHz, 1.2 W/kg SAR for two hours (pulsed or continuous), there was a substantial increase in DNA strand breaks in the brain cells after four hours of exposure.[26, 27]

The brain is the highest consumer of inhaled oxygen. Much of oxygen eventually converts to CO_2 and water. A smaller part of the oxygen forms ROS. The brain is particularly sensitive to oxidative stress because of its high metabolic rate and its composition that is rich in polyunsaturated fatty acids, an easy target for ROS!

In one study, Wistar rats were acutely exposed to extremely low frequency 50 Hz at a magnetic field intensity of 2.4 μT. The findings included impairment of catalase activity, but when this exposure was combined with radio frequency exposure, reduction of glutathione was noted as well, which clearly favours the free radical argument.[28]

An important study reported in 2001 that a 50 Hz magnetic field of 1 mT could induce micronuclei in Syrian hamster embryo cells and thus be responsible for genomic instability and indeed predispose to carcinogenicity. The authors state that it is due to free radicals and/or to unscheduled 'switching on' of signal transduction pathways.[29]

What about RF radiation from mobile phones? Ilhan et al.[30] reported increased levels of free radicals in rats—increase in free radicals, such as NO, MDA (malondialdehyde), XO (xanthine oxidase)—in the brain and decreased levels of superoxide dismutase SOD and glutathione peroxidase (GSH-Px) after rats were exposed to a 900 MHz mobile phone at a SAR of 2W/kg one hour a day for seven days. They further showed that pretreating rats with antioxidant from *Ginkgo biloba* could block such effects. They also mentioned histopathological evidence of brain injury in rats unprotected by *Gingko biloba*.

In 2015, Antonio et al. demonstrated that ELF EMFs can actually inhibit antioxidant systems—namely, catalase, cytochrome 450 and NO synthase—in erythro-leukaemic cells, thus upsetting the redox balance at cellular level, which is crucial to maintaining normal physiology and may thus have some role as well in the causation of neurodegenerative diseases, cancer, and ageing.[31]

Chapter 11

Leaky Blood–Brain Barrier

Coursing through superhighways of the circulatory system, blood is like an ocean where all rivers eventually meet. While it is the engine that transports nutrients, hormones, and a lot of messenger molecules, it is also the final destination of various toxins awaiting elimination. In order that undesirable molecules do not spill over into circulation of vital areas of the body, nature has created remarkable structural barriers between general circulation and circulation within those vital organs. The most important one, of course, is the brain—but also the gut, eyes, ovaries, and testes.

The Brain Is Privileged

The human brain, being the master control centre, is exclusively privileged by design to be securely wrapped in a thick skull and functionally kept safe from toxins in the blood stream by a barrier that, much like high prison walls, exists between it and blood circulation. A constant vigil is kept over the bidirectional exchange of permissible essential molecules by a barrier created by structural modification of cells on the inside of capillaries of the brain, which have tight junctions between themselves, with no openings and reinforced by connective tissue cells (glial astrocytes) surrounding the bilayered basement membrane.

Astrocytes on the outer surface of endothelial cells have protrusions called end feet, which help in functional regulation and repair of the blood–brain barrier. Glial cells are non-neuronal cells that maintain homeostasis, form myelin, and provide support and protection for neurons. The number of mitochondria in endothelial cells is five times higher compared to muscle cells. One in four cells covering the capillary surface is a pericyte, a type of immune cell involved in immune mechanisms protecting the brain. Such a barrier ensures that proteins cannot pass into brain tissue from general circulation.

Water, most lipid-soluble molecules, oxygen and carbon dioxide, steroid hormones, certain amino acids, and sugars can diffuse from the blood to the nerve cells. The barrier is slightly permeable to ions such as sodium, potassium, and chloride, but large molecules, such as proteins and most water-soluble chemicals, only pass poorly. However, when this barrier is damaged in conditions such as tumours, infarcts result from compromised blood flow or infections; also the normally excluded molecules can pass through, possibly bringing toxic molecules into the brain tissue.

As BBB is hydrophobic, lipid-soluble solutes can normally get through. A dysfunctional BBB allows hydrophilic molecules to pass into brain tissue. This can lead to swelling in the brain, increased pressure within the brain, and its coverings and even irreversible brain damage. Epileptic seizures and extreme hypertension can also disrupt BBB.

It is now established that BBB cannot protect the brain against radio frequency microwaves. Parathath et al. have shown that oxidative stress can cause increased permeability of the blood–brain barrier.[1]

The hippocampus resembles a seahorse with the forelegs of a horse and the tail of a fish. Sandwiched between the brain stem and cortex, it is part of the limbic system. It places priority on incoming information through the senses via the cortex of brain before the information is converted into memory and stored back in the cortex as a permanent file. Thus, it plays an important role in learning. If the blood–brain barrier is compromised, the hippocampus becomes the easiest target for toxins.

Frey Effect

Ever since the invention of the radar, a microwave-based technology, concerns were being raised about its health effects. In a strange episode reported in the 1970s, it was revealed that during the Cold War microwaves were beamed at the US embassy building in Moscow in 1953 in a covert manner. Subsequently, further research was recommended by the US Department of Naval Research and US Army to learn more about effects of MW radiation.

One of the effects of microwaves came to light by sheer accident. Long before the advent of mobile phones, in 1960 a technician whose job it was to measure signals emanating from a radar station claimed that he could 'hear' the signals directly if he stood next to the radar beam. He called up his superior, Allan Frey, then a young biophysicist at the Cornell University, who was the finest researcher in biological effects of microwave radiation at that time. Frey stood next to the beam, and lo and behold, he could hear it too! It went *zip . . . zip . . . zip.*

Frey went on to conduct several experiments on this exciting finding. He proved that even the deaf could hear it, which meant that it had nothing to do with how we hear normally. Somehow, the brain perceived it directly. It came to be known as microwave hearing or the Frey effect.

Subsequently, in one of his several experiments, he injected a fluorescent dye in the blood of rats and exposed their bodies to microwaves, and in a matter of minutes, the dye could be found in the brains of the rats. Allan Frey reported leakage of fluorescein after thirty minutes of pulsed and continuous exposure.[2] Two years later, passage of C-mannitol, inulin, and dextran was reported at very low levels.[3]

The story of the initial discovery that MW could make BBB permeable is rather unusual. In the 1970s, one of the problems in treating fatal brain tumours was in getting the drugs into blood circulation of the brain because BBB wouldn't allow that to happen. Indeed, the search was on at that time on how to deliberately make BBB permeable. That BBB could be made leaky with microwaves came as a pleasant surprise, but of course, using microwaves in an uncontrolled manner would neither be ethical nor feasible.

In its original use, microwave technology was developed by military industrial complex for covert operations. Its first civilian use was for the development of the microwave oven in 1947. In the late 1980s, when cellular phones first became available, interested scientists would look upon it as a fascinating tool for experimental work. One could now study what a controlled exposure of microwaves could do to living tissues. The pioneering work of Allan Frey was subsequently continued by a group of neuroscientists at Lund University in Sweden led by Dr. Leif Salford since 1988.

EMFs Do It Again

In terms of its structure and function, the blood–brain barrier in human and rat brains is almost identical. In a landmark study published in 1994, Leif Salford and his colleagues demonstrated that exposure to EMF at 915 MHz in continuous as well as pulsed mode for as little as an hour leads to the extravasation of albumin across BBB in 30% of rats as against only 8% in controls.[4] A normal brain would not allow albumin to sneak into its circulation because it is neurotoxic. If a molecule as large as albumin can pass through a leaky BBB, then many smaller toxic molecules can do it better. In a subsequent study, they showed that increased permeability is observed for as long as 7 and 14 days after a brief 2-hour exposure.[5]

Since the 1990s, Salford's group have used mobile phones as the source of EMFs in their experiments with thousands of rats, using TEM (transverse electromagnetic transmission line chambers) cells, which have two compartments—one above the other with enough space for the animal to move around in order to exclude stress of immobility. In 2001 Salford called the voluntary exposure of the brain to microwaves from handheld mobile phones by (then) one-fourth of world's population as 'the largest human biologic experiment ever'. In 2003 they expanded the scope of their search to find if the leaky BBB would lead to morbid changes in the brain. They selected rats 12–26 weeks old because their brains are comparable with that of teenagers.

Societal concern for RF EMF exposure is even greater in the case of young teenagers because their neurophysiological maturation is at risk. Using GSM mobile phones, they used peak power of 10 mW, 100 mW, and 1,000 mW, which exposed rats to 0.24, 2.4, and 24 mW/kg, resulting in a whole-body SAR of 2 mW/kg, 20 mW/kg and 200 mW/kg respectively only for 2 hours. The animals were allowed to survive for 50 days, at the end of which they were sacrificed and their brains removed for study. Exposed animals showed clear evidence of serious neuronal injury in the form of dark neurons.[6]

In order to evaluate how long-term use of cell phones would impact cognitive function, Salford's group have evaluated cognition in rats after a 2-hour-per-week exposure for 13 months and found significant decline in cognitive function.[7]

Paradoxical though it may appear, Persson showed that leakage of BBB is more pronounced at lower-power densities (SAR less than 2 mW/kg) rather than at higher ones (SAR 200–2,000 mW/kg). This new knowledge can be used as a foundation for new exposure limits that take into account non-thermal biological effects of microwave radiation from mobile telephones and base stations.[8]

In real-time exposure in humans, it has been suggested that the spinal cord is essentially a linear conducting structure of fair length, like a fractal antenna that can generate a significant voltage from an incident EMF wave and then leak the energy into the base of the brain, causing leaky BBB.[9]

Further research has focused on molecular mechanisms involved in causing increased BBB permeability. Almost four decades after the pioneering work of Allan Frey, J. Tang et al. in 2015 have shown beyond a shred of doubt that

900 MHz EMF fields at as low SAR as 0.016 W/kg to the whole body and 2 W/kg to the head region opens up the BBB and leads to deposit of albumin in the hippocampus and cortex, causing them to swell up, and that this happens via activation of signalling cascade MAPK-1/ERK pathway. This would explain the cognitive decline reported earlier. The MAPK/ERK (mitogen-activated protein kinases / extracellular signal-regulated kinases) pathway is a chain of proteins in the cell that carries

a signal from a receptor on the cell surface to the DNA in the nucleus of the cell.[10]

In another study, irregular firing patterns in the hippocampal neurons were shown via implanted microelectrodes in the hippocampal region in rats, which reflected an altered physiological pattern when rats were exposed to 916 MHz EMF.[11]

O. Bas and colleagues studied the impact of 900 MHz EMF exposure at 0.016 W/kg (whole body) and 2 W/kg (head region) in 16-week-old baby rats for 1 hour a day for 28 days. At the end of 28 days, the rats were sacrificed, and the growth of pyramidal cells in the hippocampus was compared with unexposed rats. A statistically significant difference was found as the number of cells in the exposed group had grown much less than in unexposed rats.[12]

In an in vitro study, C. Chen and colleagues showed that when embryonic neural stem cells (eNSC) are exposed to 1,800 MHz, apoptosis, proliferation, cell cycle, and mRNA expression of related genes are unaffected; however, outgrowth of neurites is certainly affected. This has adverse influence on development of the brain.[13] Another study supports these findings and goes a step further in elucidating changes at molecular level caused by EMFs during eNSCs differentiation.[14]

In treatment of acquired immunodeficiency syndrome, getting an antiretroviral drug (such as saquinavir) across the blood–brain barrier into intracranial circulation has always been a problem. A clever application has been found to solve this problem by combining EMF with solid lipid nanoparticles.[15, 16]

It is extremely disturbing to learn that radiation from a cell phone placed next to the head can damage the blood–brain barrier in a matter of minutes and at exposure levels, which are thousands of times less than what is regarded as safe levels of exposures by ICNIRP and FCC and IEEE.

The implication of Salford's work is terrifying when you consider that virtually everybody on the planet has a cell phone of his or her own. What does it have to do with brain cancer is yet to be established conclusively, but it certainly would predispose individuals to early onset of dementia. It is of particularly increasing concern in the case of children, who have been known to develop what is called as digital dementia!

Chapter 12

Mobile Phones and Infertility

Miniaturization of technology has allowed the mobile phones to be small enough to be conveniently carried in pockets of trousers. If you are a male teenager, that kind of convenience may come at a price. The skin of the scrotum that wraps protection around testes is so sensitive that even tight clothing can affect an extremely important process going on in there, namely the production of sperms. Male sperm is the fountainhead of life. The presence of a source of radio frequency electromagnetic radiation within inches of such a delicate process has caused a great deal of concern. Currently, scientific research points in the direction of nothing but harm that can be expected to arise out of such a beast in the vicinity of the testes.

Seeds for certain disorders that children suffer are sometimes sown in the defective sperm. It appears the children are condemned to carry the burden of the misdoings of the generation before them. While smoking, alcohol, heavy metals (such as lead and cadmium), bisphenol, insecticides, pesticides, and spermicidal foods have been known to be toxic to human sperm, there is increasing evidence of both ELF and RF EMF being just as toxic.

Sperms Are Sprinters

One out of 20 men is said to have some degree of male infertility. In almost half of couples unable to have a baby, the problem may be with the father. Each milliliter of normal semen can carry up to 15 million sperms. A sperm is a highly differentiated cell structurally designed with its head and a tail to succeed as a sprinter! It is not so much the number of sperms that counts. It takes only one sperm to fertilize the egg. If a defective sperm succeeds in fertilizing the egg, it may be responsible for early miscarriages and babies born with congenital defects, autism, epilepsy, schizophrenia, bipolar disorder, cleft palate, diaphragmatic hernia, and even childhood cancer! It is the functional viability of the sperm that matters.

So what goes wrong with the functional viability of the sperm?

Oxidative Stress Again

Aitken, in 1987, suggested that in the defective sperm function, oxidative stress was the underlying mechanism, a theory that received tremendous experimental support in subsequent studies.[1] Back again in 2014, Aitken reaffirmed his view that oxidative stress indeed underpins not only structural damage to sperm DNA but also their ability to fertilize the egg and that this is due to simultaneous damage to plasma membrane of the sperm.[2]

Two hallmarks of oxidative stress upon the sperm is the disruption of DNA and lipids and proteins in its plasma membrane. No matter if it was the genetic, lifestyle, or an environmental cause that triggered it, at the end, it is the build-up of oxidative stress at the molecular level that throws a spanner into things. The evidence for this comes from the fact that DNA damage in the nucleus of the sperm is a frequent finding that disrupts its potential to fertilize the egg.[3]

Damage to the DNA of the sperm has been known to be associated with poor semen quality, poor chances of fertilizing the egg, and increased risk of abortion. The production of ROS, mainly superoxide, by sperms is a normal by-product of their metabolism, and normally, their concentration, which is physiologically essential in empowering sperms, is low and controlled. This is possible due to normal activity of free radical scavengers, such as superoxide dismutase, catalase, and glutathione peroxidase. Thus, a fine balance is maintained. The trouble starts when the production of superoxide overwhelms the capacity of antioxidant enzymes to neutralize them. When the build-up of ROS exceeds that of available natural antioxidants, the damage occurs.[4,5]

As in the case of brain, the source for these ROS is thought to be the mitochondria.[6,7] Phospholipids that make up the cell membranes are negatively charged. Left to themselves, they would repel each other and rip the membrane apart. To overcome that, positively charged calcium ions sit between them, providing adhesiveness and stability to the membrane in addition to being important intracellular messengers.

At a purely physical level, the effect of interaction between EMF and biological tissues may disrupt the orderly fashion in which electrons are arranged in the orbits of atoms and chemical bonds between atoms of molecules. At the biological level, the disarray in electrons has real consequences, such as protein misfolding, breaks in DNA and build-up of calcium ions within cells, causing generation of free radicals with far-reaching consequences. It should be obvious in this context that broadly the effects would depend on the frequency, power density, and duration of exposure, though there are other subtle parameters that influence the outcome.

Two major meta-analytical studies appeared in 2014. The first, which was by Liu et al.,[8] analyzed the association between mobile phone use and quality of semen after reviewing eighteen studies with 3,947 men and 186 rats, and the in vitro studies clearly indicated adverse impact of RF EMFs on semen quality. The other meta-analysis, which was by Adams et al.,[9] included 10 studies and 1,492 samples and showed adverse impacts both in in vitro as well as in vivo. Another study published in 2012 by S. La Vignera et al. indicated similar results.[10]

Because the testis is a superficial organ covered only by skin, it may absorb more electromagnetic energy than other organs. Human testes need physiological temperature 2 °C lower than the body temperature for optimal spermatogenesis, and an elevation of testicular temperature may have a detrimental effect on sperm production. Use of tight clothing in the region of the testes may make it warmer than the rest of the body. It is estimated that only a SAR value greater than 4 W/kg could result in a temperature increase of 1°C. Therefore, at this time, there is no clear-cut evidence that supports the thermal effect of mobile phone radiation on the human body.

Pulsed RF EMFs generate a whirlpool of currents that shake calcium ions on and off the membranes. A defective membrane of the sperm will not easily allow it to fuse with the vitelline membrane of the ovum. As the calcium homeostasis is disrupted, there is intracellular build-up of enormous amount of calcium, and that is the beginning of the disaster, namely, the generation of unmanageable numbers of free radicals.

Through *in vitro* and *in vivo* studies, EMF exposure has been found to alter the reproductive endocrine hormones, gonadal function, embryonic development, pregnancy, and fetal development. Alterations in biomarkers following EMF exposure have been reported through *in vitro* and *in vivo* experiments using animal cells and animals, respectively.

Decrease in sperm motility and viability and increase in ROS and decrease in ROS-TAC score in the exposed group of neat semen samples have been reported by Agarwal et al.[11, 12]

Sperms have polyunsaturated fatty acids in their membranes and limited antioxidants. ROS also has damaging influence on DNA enzymes, lipids, and proteins. Chronic exposure has been linked to decrease in activity of catalase, SOD glutathione peroxidase, the antioxidant enzymes in various organs of the body.

In a study by Adel Zalata et al.,[13] the authors have focused on the effects of RF on several parameters of the human sperm—namely, motility, fragmentation of DNA, and expression of a gene in semen called CLU (clusterin)—on 124 semen samples and found statistically significant decrease in sperm motility, linear velocity, linearity, and acrosin activity and also increase in DNA fragmentation and CLU gene expression, which is a marker of cell under stress as also shown by Strocchi and colleagues.[14] CLU has antioxidant properties and attempts to protect cells from free radical damage, and its overexpression is an indirect evidence of increased oxidative stress.

S. Kumar et al. have further shown loss of sperm count and oxidative damage, causing not only DNA damage but physical reduction in seminiferous tubules, thereby reducing testicular weight besides affecting fertility in male albino rats.[15]

Not only RF EMF, even 50 Hz ELF EMF have also been shown to impact the motility of sperms and overall fertility.[16, 17]

Gorpinchenko et al., in their study reported in 2014, have shown that exposure of semen of healthy males to mobile phone in 900/1,800 MHz radio frequency for five hours, kept at a distance of 5 cm in standby/talk mode, and with a call being made to it once every ten minutes causes significant decrease in sperm with progressive movement, increase in sperm with non-progressive movement, and evidence of DNA fragmentation.[18] In one study, successive generations of mice were exposed to EMFs. They had to stop at the fifth generation.There was no sixth generation as the fifth generation had lost the ability to conceive.

Erectile dysfunction as a result of carrying mobile phones in the pant pockets has also been found as a result of wireless RF radiation.[19]

Whittow et al. showed that if a mobile phone is kept in a trouser front pocket, common objects like coins, rings, or zips in pockets of trousers increase the absorption of RF energy in the area of the waist above the levels considered safe in United Kingdom.[20]

The use of a laptop connected to the Internet via Wi-Fi and positioned near gonads may affect the sperm.[21] Healthy sperm stays fertile in the female reproductive system up to 80 hours after sexual intercourse. Therefore, the use of laptops in women trying to conceive can be hazardous.[22] Besides, once conception has occurred successfully, there is evidence that microwaves can reduce blood flow in the placenta.[23] That is certainly something to think about if you are a prospective mother!

Chapter 13

Heat Shock Proteins

Just as we are susceptible to emotional stress, our cells are vulnerable to stress as well, such as stress of physical injury, environmental toxins, or extremes of temperature. Through molecular mechanisms that elaborate certain stress proteins, cells have a short-term response to minimize acute damage as well as a long-term adaptive response to learn to deal better with similar insults in the future.

Feruccio Ritosa, a pioneer of genetics in Italy, was fascinated with fruit fly as an experimental subject. He regarded it as a species 'between bacteria and man'. While studying chromosomal activity of cells, one of his colleagues had accidentally changed the temperature conditions, and Ritossa noticed a different pattern of chromosomal puffing. This led to the discovery of new proteins. He named them heat shock proteins.[1]

Now we know that they are expressed also in response to extreme cold, ultraviolet light, and inflammation due to infections, toxins, and starvation, etc. They exist inside cells of all living forms. Based on the molecular weight, there are various families of heat shock proteins—namely, HSP 20, 27, 60, 70, and 80.

In order for them to be functionally active, proteins need to configure themselves to a new shape. The change from a coiled shape to a three-dimensional structure occurs as a result of protein folding as determined by the sequential arrangement of amino acids. Errors in folding can have disastrous consequences, such as neurodegenerative diseases. HSPs play a role in appropriate protein folding and in preventing undesirable aggregation of amino acid chains. HSPs can refold misfolded proteins, thus modulating immune response to arrest inflammation.[2] HSPs also confer protection to the integrity of the blood–brain barrier when the brain is threatened by compromised supply of blood to the brain.[3]

Now that nonionizing electromagnetic radiation is a known environmental stressor, its effects on HSPs have been studied. Lin et al. showed in 1999 that the induction of HSP 70 in response to 60 Hz fields[4] and the same group showed that exposure to ELFs (<300 Hz) induces HSP 70 proteins.[5]

In 2010 Mannerling and his group have reported that magnetic component of EMRs induce HSP 70 in human leukaemia cell line K562, an effect that can be blocked by employing an antioxidant such as melatonin.[6]

A recent study has shown that in personnel working in information technology, along with HSP 70, C-reactive protein levels in the blood are also raised, indicating unnoticed inflammation in the body.[7]

Expression of HSP 70 has been demonstrated in the hippocampus of rats, indicating a stress response in the brain in response to EMFs.[8] This is highly significant as the hippocampus plays a major role in memory and learning.

If exposure is chronic, it leads to activation of HSP 70 as a protective response. Normally, HSP form a protective layer around enzymes to protect them, but their overexpression is a significant sign of cells being under stress. With high levels of HSP, the blood–testes barrier's permeability is altered, exposing the sperm to toxic influences from general circulation and thus affecting fertility.

Chapter 14

Radio Frequency and Wildlife

> If the bee disappeared off the surface of the globe then man would have only four years of life left. No more bees, no more pollination, no more plants, no more animals, no more man.
>
> Albert Einstein

In a large part, we owe our understanding of effects low frequency power lines and radio frequency on human health to the animals sacrificed as surrogates to predict harm to humans. FCC and ICNIRP standards, howsoever fallacious, serve as a benchmark for safe limits of exposure. Ungrateful as we are as a race, there are no standards as yet devoted to the safety of animals and plants.

Birds Never Get Lost

Migratory birds have been known to use environmental cues, such as geomagnetic field lines, light, and position of the sun and stars along their long journeys. They have no GPS or satellite navigation systems yet have never been known to get lost. In the 1960s, the red-breasted robin was the first subject of study of magnetoreception in birds when it was discovered that animals sense the direction of geomagnetic field and use it as a compass for navigating around as well as during migration. The robin was named the national bird of United Kingdom in June 2015. Various other experiments have proved it to be true. There are two theories about how they achieve this magnificent feat—namely, a radical pair theory and a magnetite theory. The former works in the presence of light and the latter even in its absence.

The sensory perception of magnetic field lies in the eye of the bird. A bird requires the presence of blue wavelength of light to be able to sense it. The radical pair model of magnetoreception suggests that the initial step is the absorption of photon of light by cryptochrome, a molecule in the blue-light photoreceptor in birds' eyes; in turn, it forms a pair of molecules with unpaired electrons (radical pairs), which are sensitive to inclination of geomagnetic field lines.[1] The resonance thus created allows birds to use geomagnetic field to navigate. Identical mechanisms have been found to exist in other species, such as fruit flies, insects, lobsters, honeybees, and aquatic animals, such as salmon and turtles and even in plants.

In the absence of light, however, ferromagnetic resonance is brought about by sensory perception of geomagnetic field by iron-containing magnetite in the beak of the bird.[2] The latter can be tested. If the beak of the bird is anesthetized, they get disoriented in the dark. The two theories are complimentary and not necessarily in conflict.

In 2004 Ritz showed that if the geomagnetic field lines are superimposed with either a broadband (0.1–10 MHz) or a single 7 MHz oscillating electromagnetic field, interesting things happen. If the superimposed field is aligned with the geomagnetic field, nothing changes, but if it is imposed at an angle of 24° to 48°, birds get disoriented.[3]

In the wake of industrialization of our societies, frequent incidents of electrocution and collisions were bad enough for birds, but now we have invisible radio frequency electromagnetic radiation affecting their survival. In 2014 Engel and his group have shown that when exposed to man-made electromagnetic radiations at levels below WHO guidelines, European robins lose their magnetic compass.[4]

In a major meta-analysis reported as early as 2013 on effects of RF EMF on birds, honeybees, fruit flies, mammals, and plants, ecologically relevant evidence has been found in 50% of animal studies and 90% of plant studies.[5]

What if Pollination Didn't Happen?

Bees are a very important part of nature's food production machine. Honeybees are invaluable not only for the honey and beeswax that they produce but also for pollination of various crops. It has been estimated that 35% of production of fruits and vegetables is dependent upon honeybees globally.[6]

Bees make use of natural magnetic fields for their survival. It is through these fields that they learn to orient themselves to their surroundings, to communicate with one another, as well as to anticipate changes in weather. In the past, beehives were common sights even in urban areas but not any more. Several observations suggest that man-made electromagnetic fields disrupt their survival strategies. They fail to find their way home in the presence of superimposed man-made fields.

Throughout the history of beekeeping, seasonal variations in their colonies are a normal phenomenon. A strange behaviour was reported in 2006 when large colonies of honeybees abruptly began to disappear in North America and did not return, leading to serious economic losses.[7] No ill or dead bees were ever found in or near the beehives as would occur if it were pesticides, varroa mites, GMOs, or very harsh winter. The disappearance act of bees in unprecedented numbers corresponds with increase in installation of cell phone towers.

This has been termed as colony collapse disorder (CCD). Curiously enough, it was in the same year that the transmission power of HAARP[‡‡] was nearly quadrupled from 960,000 watts to 3,600,000 watts. Interesting coincidence. Similar phenomena were noticed in Western Europe. Scientists have observed that high-tension power lines[8] as well as exposure to RF EMF at power densities as low as 8.549 $\mu W/cm^2$ from 900

‡‡ HAARP (High Frequency Active Auroral Research Project) was launched by US Air Force and Navy in 1993 in Alaska. It employed a very powerful radio frequency directed at a part of the ionosphere allegedly to study potential for radio communication and surveillance. The excited portion of the ionosphere caused superimposition of unnatural magnetic fields to several thousands of nanotesla that might have been responsible for the disappearance of bees. The project was shut down in 2014.

MHz GSM mobile phones adversely affects colonies of bees.[9] Predictions are that in the United States, honeybees will be extinct by 2035 and even sooner in the United Kingdom. Science has not yet developed a technical alternative to natural pollination, but when that becomes necessary, it might cost a million times more than a nation's GDP.

Geomagnetism Guides Animals' Movements

The abilities of animals to anticipate weather and to run for cover long before the arrival of hostile conditions are well documented since long ago. Hours before a devastating tsunami hit the coast of Thailand in 2004, animals were reported to display a strange behaviour. They were running in the opposite direction even though their masters were trying to herd them towards the beach.

The fish, stingray, sharks, and whales do not swim about in the sea randomly or purposelessly. Insects rely on multiple factors for their navigation and orientation, such as colours, aroma in the air around them, gravitation, and even voltage in the air.

Termites build large mounds in north–south axis. Dogs are known to go around in circles before they poop. That's not without a purpose. They need to align themselves with magnetic field lines.[10] This opens up fascinating avenues for further exploration in the relationship between animal behaviour and geomagnetism as dogs are widely used as experimental animals.

A definite decline in the breeding, nesting, and roosting of house sparrows (*Passer domesticus*) has been reported in several countries. This has been linked to cell tower base stations.[11] House sparrows in the wild are driven away from their natural habitat if masts are installed in their environment.[12] Chicken embryos have been reported to die of heart attack, resulting from exposure to 915 MHz cellular phone radiation,[13] and similar results have been replicated by other researchers.[14, 15]

Marine mammals, fish, and sea turtles are also sensitive to electromagnetic environment. Although transmission electric cables have been laid in the seabed since nineteenth century, there is only sparse information available on their impact on marine life. Marine species depend on their magnetic sense for orientation in their environment and for short- and long-distance navigation and on their electric sense for detecting a prey as well as avoiding a predator. Cartilaginous fish, such as sharks, skates, and rays possess under their skin a tiny canal connected to their brains, known as ampullae of Lorenzini, which are electrosensitive and can detect weak electric cues produced by the prey.

EMFs generated by the undersea telephone and communication cables and induced heating of oil and gas pipelines would be expected to alter background geomagnetic field and interfere with these functions. Sea turtles navigate from their current location towards the target location, where they expect to find food with the help of their magnetic sense as one of the major cues.[16]

Balmori has extensively studied impact of electromagnetic radiation on plant life. In a study from Spain, he revealed that poplar and willow trees show consistent pattern of deterioration when exposed to electromagnetic fields. They do not attain their expected height, the treetops appear dry, and they have high susceptibility to diseases.[17]

Not only animals, even plants express cellular stress response when exposed to 900 MHz electromagnetic field within minutes after exposure.[18] No wonder then that radio frequency radiation can retard germination, as has been shown in wheat plants.[19]

Extinction of Species

In a 2014 report of the Living Planet Index of Zoological Society of London, populations of mammals, reptiles, fish, birds, and amphibians have declined by an average of 52%. The worst impact is on freshwater species, which is 76%. In the last 40 years, the overall population of animal species has halved!

The index keeps track of 10,000 species worldwide since 1970. Nature is unable to cope with the extent of human activity. If you cut down one tree, it takes a few years to get its replacement—that is, if you plant one. Fish reproduction is falling short of human consumption. Carbon emissions exceed the ability of forests and oceans and atmosphere to absorb them without tipping the balance towards harm. How much it has to do with EMF pollution is an open question.

Chapter 15

Carbon Footprint of Wireless Technology

Everything is a poison. It is only a question of dose

Paracelsus (1493–1541)

Anthropogenic carbon footprint is greenhouse emission caused by use of energy to produce goods or services by an individual, organization, or nation.

Classically, the use of electrical energy has been for illumination, cooling and heating, transportation, and running electric motors of enormous kinds in a variety of appliances and machines. Each one of these has a carbon footprint of its own. The emergence of wireless radio frequency–based devices has created a demand for the use of electric energy unlike ever before. Largely invisible to the public eye, the use of energy to run the information and communications technology (ICT) ecosystem is exceptional in that it is a 24/7 machine that never stops. It uses 1,500 TW hours of electricity annually, which is equal to what was used to illuminate the entire world in 1985. That is about the total power generated by Germany and Japan put together.

With the advent of wireless broadband highways, we are now in the zettabyte (10^{21} B) era. Annual use of electricity to support such massive transport of data bits exceeds that used by the aviation industry globally. When you watch a high-resolution video on a smartphone, consumption of energy by the battery of the smartphone may be minimal, but the invisible consumption by broadband networks between user, cell tower, and data centres that allow you to uplink and downlink is mammoth. For instance, watching a video an hour a week for a year consumes more energy than a refrigerator does over a period of two years.

Internet traffic has been on the rise in leaps and bounds since the early part of this century, and energy consumed to keep it running is unlike anything mankind has ever known. There are more bits transported each moment than passengers transported globally. According to one estimate, by the year 2030, it is likely to be twice as much as today—not to speak of energy-intensive manufacturing processes to meet massive demands for these products. Greenpeace estimates it might be three times as much.

The technological pedestal supporting wireless cloud comprises data centres connected to your PC, iPad, laptop, or smartphone through global broadband telecommunication networks. Wireless cloud has further revolutionized telecommunication.

It may be argued that the increase in cost of energy use due to wireless technology is indeed cost-effective rather than remaining dependent on older analogue technology. After all, going paperless saves trees. Trading is much faster today than ever before. Money moves instantly. There is a huge economic advantage in adopting wireless technologies. As a matter of fact, the carbon footprint of other industries is much higher than that of the wireless industry.

One may then ask, why the fuss about energy consumption and carbon footprint of wireless technology?

More Is Less

The moot point is that human appetite for wireless technology is insatiable. Preying on this appetite is the industry inventing innumerable, often preposterous, applications of wireless technology and creating a universal market of uninformed consumers. The information communication technology is the fastest-growing industry with an annual increase of 40% in Internet traffic.

Microwave radio frequency–based wireless transmission systems are used in transmission of radio, television, satellite communication, police and military radar, federal homeland security systems, emergency response networks apart from billions of personal communication devices. They consume millions of watts of energy.

Most of new phone subscribers prefer mobile phones as their primary phone. Landlines are fast getting phased out. Governments need not expand wired infrastructure in areas not having landline services as infrastructure for wireless services is both technically easier to install and cost-effective.

We Are Trapped

Typically, the average intensity of man-made radiation in urban areas today is 10,000 $\mu W/m^2$. This is 10,000,000 times more than natural background radiation and much higher than what is regarded at safe limit by governmental organizations. Smart metres send 1W output at 2.4 to 2.48 GHz range to a local access point and further to a distant information centre. The system is networked to cover entire communities through a combination of mesh-like networks called WAN (wide area network). WiMax (which began in USA in 2009) aims to roll out much more powerful connectivity capabilities and will require bringing more towers (which work at lower frequencies but with much higher power densities) in close proximity to inhabited areas.

This is unprecedented in the history of industrialization of human civilization. As end users, we fail to realize the hidden aspect of energy use and the carbon footprint it creates. The combined energy consumed by billions of devices defies estimates.

In 2012 Greenpeace evaluated the use of clean energy by various companies and gave poorest ratings to Apple, Amazon, and Microsoft, three of the biggest players. It further showed that most companies still depended on fossil fuels as a source of energy. Because of its lowest cost of production, coal continues to be the choicest source of production of energy even today. It contributes 40% of total energy production even today. Use of resources other than coal to generate such a massive demand for energy is cherishable but still a distant dream.

Based on data produced by the Earth System Research Laboratory of US Department of National Oceanic and Atmospheric Administration, the annual mean global carbon dioxide emissions increased from 0.96 ppm (parts per million) in 1959 to 1.25 ppm in 2000, an increase of

0.29 ppm in 41 years. However, in 2015 it was recorded at 2.92 ppm, an increase of 1.67 ppm in 15 years. That was the period during which wireless telecommunication industry proliferated.

The number of wireless cloud users was 42.8 million in 2009. Given the annual growth rate of 69%, it may have exceeded 1 billion. According to a report by CEET (Centre for Energy Efficient Telecommunications), the magnitude of carbon footprint caused by use of energy by wireless cloud rose from 6 Mt in 2012 to 30 Mt of CO_2, an increase of 460%. This is the same as adding 4.9 million cars on roads. Up to 90% of this consumption is attributable to wireless access network technologies, whereas data centres account for only 9%.

According to the GSM Association, 19.1 kWh energy is required to consume 1 GB of data. An average annual consumption to charge an iPhone is 3.5 kWh, and one iPhone connection costs 23.4 kWh annually. If an average user consumes 5 GB of data in a month, he or she would use 1,173 kWh of electric energy in a year. This is roughly equal to an average annual household consumption of electric energy in India and one-fourth of that in UK.

And Now IoT

The Internet of things (IoT), the next big example of a networked planet, is already here. Although still regarded to be in its plumbing stage, in the UK alone, 40 million devices are already networked. The next generation of appliances for daily use (such as microwave ovens, refrigerators, toasters, dog collars and wristbands for kids to locate them when they are outdoors, and myriads of other devices, including your furniture) will be embedded with sensors and transmitters and wirelessly connected to your mobile phone as well as the manufacturer.

You can preheat your oven or turn on the air conditioner using your mobile phone app before you arrive at home, and manufacturer can keep track of your pattern of usage and build a database for their research and design and advance their marketing strategies. This will employ communication protocols using Wi-Fi. Current estimates indicate that up to 200 billion such devices may be connected before the end of 2020.

This will undoubtedly add billions of zettabytes (1 zettabyte = 10^{21} B) of data on the already-overloaded wireless cloud.

There will be need for separate gateways dedicated to the Internet of things. Samsung has announced that in the next five years, all of its devices will be networked on IoT. General Electric already makes appliances embedded with RFID. They haven't mentioned how often they would ask their customers to buy software upgrades. The challenges arising out of the Internet of things are not yet fully known, but what is clearly known is that it will generate colossal demand for electric energy—not to mention billions of not-so-smart retired devices that would need to be inevitably either dumped into landfill or incinerated in the junkyards.

In April 2016, Paris Climate Agreement was signed by 178 countries to reduce greenhouse gas emissions. 'The world is in a race against time. The era of consumption without consequences is over,' said UN secretary general Ban Ki-moon at the signing ceremony. Until June 2016 only 19 countries had ratified the agreement. The less spoken aspect of the agreement was that it was optional and legally non-binding!

While a move towards the use of the sun and wind energy to generate electric energy is a welcome step, it is unlikely that the use of fossil fuels is going away any time soon. The sun and the wind are weather dependent and cannot be relied upon totally to support an industry that runs 24/7. This is why even though many wireless towers in the world are run on solar and wind sources of energy, they also have a diesel backup.

There is still no legislation in place mandating cutting the size of carbon footprint of this industry. The two slogans of every forward-looking government—to cut down greenhouse gas emissions on one hand and to provide global broadband wireless connectivity not only between human beings but also machines—have a hidden contradiction. It would appear impossible to achieve one without sacrificing the other. The resources of the planet are not infinite. Whether this trend is sustainable in the longer term is a huge question with no answers as yet.

Chapter 16

Politics of Precautionary Principle

A precautionary principle is a decision exercised when scientific information is insufficient, inconclusive, or uncertain and where there are indications that the possible effects of the contentious agent on the environment or human or animal or plant health may be 'potentially' dangerous and inconsistent with the chosen level of protection.

Given the immense progress science has made today, why is the question of adverse health effects or otherwise of radio frequency electromagnetic fields (RF EMF) still unanswered?

Or has the answer been found but not yet revealed?

There are excellent examples of brilliant work performed by scientific academia, such as Henry Lai, Narender Pal Singh, Lennart Hardell, Leif Salford, Franz Adlkofer, and many others. However, the regulators and policy makers, under corporate pressure, tend to take advantage of the lack of scientific consensus and lean towards poorly defined concept of weight of evidence as an alibi to deny the obvious, thus opening up space for the industry to hijack the truth. They fail to separate wheat from chaff!

Legislative bodies may be faced with a dilemma of setting standards about safety or otherwise of technical specifications of products that produce EMFs if they insist on hard-core evidence in humans. The fact of the matter is that the time lag between clinical manifestation of chronic diseases and the beginning of the disease process may be enormous, running into several years. For instance, in the case of holocaust survivors after the atomic explosion in Hiroshima and Nagasaki, it took up to 40 years in several cases before cancers were detected.

ALARA or AMARA?

Take a look at the precautionary principle advocated to reduce exposures to a level as low as 'reasonably' achievable (ALARA). It would appear that to call it ALAA (as low as achievable) would have served the purpose. Inclusion of the term *reasonably* leaves the door open for varieties of interpretation depending upon which section of stakeholders you represent. It is public posturing to safeguard the interests of the industry. Indeed, it is a consequence of such chicanery that, in practice, ALARA has become AMARA (as much as reasonably achievable).

Most people carry their mobile phones on their bodies. Women even keep them inside the bra. Teenage girls go to bed with cell phones under the pillow. They warn you to keep the cell phone at least 15 mm away from the body. You are mistaken if you think that is a polite advice for your safety. It is a legal clause for *their* safety. If you got brain cancer and went to the court, lawyers will ask you to prove that you kept your Apple iPhone at least 15 mm (25 mm for Nokia or Motorola) away from your body over a long period of use.

The literature says this device meets FCC safety standards when the device is used at a certain distance from the body, which means they know as well that the device is not safe if worn on the body. Medical insurance companies refuse to entertain claims for health problems resulting from the use of wireless radio frequency devices of any kind as a standard exclusion!

While it is true that a scientific study gains validity if its results can be replicated by another study, to say that it loses validity if another study does not replicate its results is to say that the second study is more reliable than the first one. As a matter of fact, when a second study is done with the exclusive purpose of validating the first study, there is room for fraud because an indiscernible and miniscule change in experimental conditions can skew the outcome in your favour.

Angelo G. Levis, in a study published in 2011 on the evaluation of all case control and cohort studies on biological effects of RF EMF, showed that negative results published in literature are largely seen in industry-funded studies.

Revolving-Chair Scientists

When war games are planned, the common man is the last consideration in the hierarchy of interests. You are lucky if you survived; if you didn't, the planners can indemnify themselves by calling you collateral damage. Governments have allowed corporate industries to not only fund research but also to determine the goals of research! When goals are predetermined, research is only a proxy. This has vitiated, corrupted, and polarized the scientific community.

There is a history of scientists who wear different hats in different chairs, having switched jobs between working for the industry and being on advisory expert committees who have hidden conflicts of interest. Conflict of interest disclosures depend on whether or not the discloser wants to make full disclosure. How safe is it to leave policymaking in the hands of such 'experts'? You can ask that question only if you live in a true democracy.

Even political careers are made or marred in the marketplace of RF EMF industry. The man who volunteered funding research into the safety of mobile phones on behalf of the industry was also the man who rebuffed the conclusions of that research when they turned out to be contrary to what he would have liked. His name is Tom Wheeler. Currently, he is the head of FCC, which regulates the cell phone industry in the USA.

A Lesson Learnt the Hard Way

There is a price to be paid for standing up to the industry, as many eminent scientists with impeccable credentials eventually found out. I shall quote two examples.

Henry Lai (an affable and modest professor of bioengineering at the University of Washington, Seattle) along with a co-researcher, Narender Pal Singh (a 1972 graduate from King George Medical College, Lucknow, India), did some foundational work in mid 1990s on the effect of electromagnetic fields on the DNA of rats and clearly demonstrated that these fields caused single- and double-strand DNA breaks. They are widely quoted throughout medical literature on the subject. The National Institute for Health (NIH) financed the project. NIH received a complaint that Lai's work was beyond the scope of the planned studies. Lai explained himself, and the matter rested there.

The complainant, as was found out later, was Bill Guy, a colleague and a co-author of several publications of Lai. At the time, their group had close academic collaboration with scientists at Motorola, the leading mobile phone company that had launched their shoe phone a few years earlier. Motorola were seriously upset with the findings of Lai and Singh. Motorola drafted a plan to war-game the duo and discredit their findings by employing for-hire scientists to conduct orchestrated studies to disprove that microwave radiation could cause damage to the DNA of rats. Even the University of Washington president Richard McCormick was approached to fire Lai and Singh.

'We thought they were collaborating and interested in science,' Singh said.

'We were naive. This shocked me, the letter trying to discredit me, the war games memo; as a scientist doing research, I was not expecting to be involved in a political situation. It opened my eyes on how games are played in the world of business,' Lai said.

The University of Washington declined the request to fire Lai and Singh. But the impact on their collaborating scientists at other institutions was worse. Many of them lost funding and some of them even their positions.

Over the next ten years, almost an equal number of studies appeared, debunking the argument that microwave radiation caused any harm to biological tissues. While Lai and Singh did not swerve from their position on DNA damage by radio frequency, the industry succeeded in proliferating the business to an unprecedented scale behind the engineered smokescreen of doubt.

In 2006, Lai analyzed 326 studies on the biological effects of electromagnetic radiations and found a 50–50 split between for and against groups. On a second look, when he segregated them into industry-funded and independent studies, he found that only 30% of the studies in the former group showed effects, whereas in the latter group, 70% did.

'Even if you accept all the industry studies, you still end up with 50–50. How could [the other] 50 percent all be garbage?' Lai asks.

On 21 April 2015, the district court of Hamburg pronounced a verdict: 'The admissible complaint is well founded . . . the disputed remarks violate the plaintiff with regard to her common personal rights . . . deliberately untrue factual allegations and those, whose falsehood is already certain at the time being made must not be accepted.'

The verdict pertained to a lawsuit filed in August 2014 by Elisabeth Kratochvil, a technical assistant who worked for REFLEX project led by Prof. Franz Adlkofer at the Verum Foundation, Germany (2000–2004). It was against Prof. Alexander Lerchl, a biologist from Jacob University in Bremen, Germany, for accusing her of having faked the results of studies showing that microwave radiation from cellular phones caused damage to human DNA. The target was to tarnish the scientific and personal reputation of the REFLEX group.

Prof. Alexander Lerchl, who was also at the time a member of the Committee on Nonionizing Radiation in the German Commission on Radiological Protection (Strahlensschutzkommission) alleged that the data in two published studies in *Mutation Research* and *International Archives of Occupational and Environmental Health* was falsified.

Lechrl's comments referred to the chief author of the study, who published a scientific rebuttal of Lechrl's comments in the same issue of the journal.

Up until then, it appeared to be no more than a matter of scientific debate between colleagues, but the bombshell came when Lechrl decided to leak it to the press and make a public spat that smacked of envy and rivalry.

On 29 August 2008, an esteemed journal, *Science*, caused shockwaves as it ran its headlines: 'Scientific Misconduct: Fraud Charges Cast Doubt on Claims of DNA Damage from Cell Phone Fields'. It referred to charges levied by Alexander Lerchl against the REFLEX group.

Frustrated with the murky wheeling-dealing, Elisabeth Kratochvil (then Diem) resigned to pursue her studies for MBA. Prof. Lerchl, on the other hand, was promoted to be the head of the Committee on

Nonionizing Radiation in the German Commission on Radiological Protection (Strahlensschutzkommission). The time for such accusation was opportune because, at that time, the European Union was considering an application from Verum Foundation for grant of research funds to further carry out experiments to establish the effects of electromagnetic microwaves in humans. The fact that the papers had passed the test of peer review and had actually been published rang alarm bells in the industry and political circles because of the potential impact of research findings. Any further substantiation of the facts would spell disaster for financial interests of the industry. The worst, however, was yet to follow.

Prof. Lerchl, with his political clout, strategically countered REFLEX group by bringing pressure upon the rector of the Medical University of Vienna as well as the editors of the journals that had published the articles to retract the studies with or without consent of the authors. An ethics commission was set up, and in its first meeting, it hurriedly reported that the accusations of fraud were found to be correct.

Prof. Hugo Rudiger, the then director of the Department of Occupational Medicine at the Medical University of Vienna and co-author of the contentious publications, felt morally obliged to respond to the findings of the commission. Besides, he had been misled to believe that Elisabeth Kratochvil had admitted that she had fudged the data.

The writing was on the wall. Prof. Rudiger agreed to retract one of the two studies. Several days later, he discovered that the chair at the ethics commission was a lawyer who had been on the payroll of a telecom company! Meanwhile, he was also convinced that Elisabeth had done no wrongdoing. Supported by Franz Adlkofer, he dug in his heels and refused to retract the paper.

Two subsequent ethics panels exonerated the authors of any misconduct. Five years later, the district court of Hamburg also pronounced a verdict in favour of the plaintiffs.

Seven years later, in 2015, in one of his own published studies, Lerchl admitted that his own research group had found increased risk of tumours of the liver and the lung in mice with exposures of 0.04 and 0.4 W/kg SAR, which is well below what is regarded as safe limits.

It Is Politics after All

Notwithstanding its ostentatious political morality, the facts (and effects) of exercising precautionary principles are as follows:

- It absolves the governments of their political responsibility. (Well, you were warned!)
- It protects the industry from being held responsible and liable.
- It allows the industry to surreptitiously continue deploying wireless devices and infrastructure.
- It mandates more funds for research.

Furthermore, it creates new jobs in research and technology.

When it was finally learnt that the mosquito causes malaria, there was no further money in the malaria story.

Does that ring a bell?

Chapter 17

Can We Really Protect Ourselves?

There has never been a technology in the entire history of mankind that has been adopted as pervasively as wireless technology. At the time of this writing, according to GSMA Intelligence real-time tracker, the number of mobile smartphones and not-so-smartphones and M2M devices in the world stood at 7.8 billion, which exceeds human population by far. As in their population, China and India top the list. When you consider the might of the trillion-dollar industry with just one product that rose in its sales from 0 to 7.8 billion in just three decades, you have a fair idea of what you are up against. A technology that you can carry in your pocket, wear on your wrist, or hook up on your ear should obviously raise a concern over what it is doing to your body.

What with a worldwide meshwork of electromagnetic waves this technology rides on, can you possibly protect yourself from something that is as ubiquitous as air itself?

In the post-industrial era, the evaluation and management of air quality in order to protect human health is a mandatory obligation of the governments. Air quality measurement stations, regulated by governments, evaluate pollutants such as carbon monoxide, nitrogen dioxide, ozone, suspended particulate matter, sulphur dioxide, lead, and hydrogen sulfide. Electromagnetic pollution is not yet on the list. While there is a large body of evidence of harm from exposure to mobile phones, cordless phones, or tablets, there isn't sufficient data on exposure from phone masts because the latter is extremely difficult to quantify due to the enormous number of variables involved.

It is not easy to characterize exposure to ELF EMF. Exposures are complex and occur at home and at workplaces and from several sources. Occupational exposures in workplaces with higher fields differ from exposures to general public. Occupational exposure is a consequence of employment. Subjects are usually otherwise healthy adults. The advantage is that the exposed personnel are made aware of risks of exposure and assisted with the need for taking precautions through administrative and technical means. The general population comprises a cross section of all ages and includes the elderly and the children and pregnant women who may not be aware of the potential for exposure and thus may not be in a position to take precautions.

Every country has its licensing body to authorize devices that generate RF and MW radiation for the purpose of radio and television broadcast, cellular telecommunication, paging, and ground-based dish antennas of satellite–earth stations, including personal devices that are actually used by individuals. However, FCC in North America and ICNIRP in European Union and United Kingdom were the first to set standards, and their recommendations are the basis of standards in other countries.

Even so, guidelines are not uniform throughout the world. For instance, in Russia, guidelines are more restrictive than in North America and some parts of Europe. EU guidelines are implemented uniformly in most member states of EU. In Luxembourg, guidelines are perhaps most strict, followed by Italy.

Exposure to ELF band of EMF and RF MW band of EMF are two distinct sets of problems because they may have different interactions with living systems.

Power frequency ELF fields are defined as electric field strength (V/m) and magnetic flux density (µT), whereas in the case of RF MW, safety limits are additionally defined in terms of power density (W/m^2).

See Appendix III to X. See Appendix XIV for magnetic fields from home appliances and certain occupations.

Acceptable power frequency electric fields should be no more than 0.3 V/m and magnetic fields no more than 0.2 µT (2 mG). Radio frequency electromagnetic fields should not exceed 1 µW/m^2. But these are utopian figures. In real life, electric fields inside homes can vary from 0 to 10 V/m. GSM mobile phones and DECT phones produce magnetic fields up to 20 mG near the head and 10 mG at a distance of 7.5 cm.

Residential wiring, heating, cooling, and use of appliances cause exposure to magnetic fields. Exposure is much higher in the case of appliances used in close contact with the body, such as hair dryers, shavers, electric toothbrushes, and electric blankets. ELF range magnetic fields are more important for biological effects, and they travel unhindered by wood or brick wall. You do not have to touch a live wire to get affected by magnetic fields.

According to WHO, exposure to magnetic fields exceeding 4 mG increases the risk of childhood leukaemia twofold. *Bioinitiative Report* indicated that risk rises substantially above 2 mG. A study from McGill University in Canada puts the safety limit at 0.1 mG. For a contrast, a microwave oven emits a magnetic field of 300 mG, and a vacuum cleaner emits a field of 700 mG (see Appendix XIV). The worst time may be at night when you spend long hours in one place. For instance, while sleeping next to electric devices that are either on or on standby mode during the night, one may get peak exposure. It is possible to calculate exposure to a magnetic field for a person by having him wear a magnetic field monitor, the same as a holter monitor, for 24 hours or use a stationary monitor inside house or workplace and calculate its average. Spot measurements are unhelpful.

The omnipresence of ELF EMFs virtually precludes any guarantees of safety against them. Worse still, there are no international standards on electric and magnetic fields produced by appliances. However, what matters is the strength of the magnetic field at the source, your distance from it, and the time you spend near it. Generally, you should be at an arm's length from the source. You may need to rearrange the ergonomics of your position and that of the sources around you. A distance of at least 2 ft between the bed and the wall behind it may reduce overnight exposure to magnetic fields significantly. Unplugging all non-essential appliances at night would help reduce the exposure.

In the case of RF MW range of EMF, risks appear to be even higher as there is no way to keep them at an arm's length. It was hoped that setting the SAR limit to manufacturing of cellular phones would solve the problem. It continues to be the industry standard and touted as 'in compliance with guidelines'. But as it turns out, it was mere hogwash. To exercise precautionary measures in the use of personal devices is too simplistic. We are constantly surrounded. It is difficult to quantify such exposure as they occur from cell towers, television towers, radars, myriads of routers in homes and workplaces and community places.

You walk into an urban building and turn on Wi-Fi, and you will find tens of hundreds of available networks. Even if you have your mobile phone on airplane mode, you are still exposed to far-field radiation. Exposure immediately beneath the antenna may actually be minimal because the radiated beam from the antenna is not angulated vertically downwards. Therefore, if you have an antenna on your rooftop, removing it is not the solution to get rid of radiation.

The truth of the matter is, unless a radical change occurs in the technological design (of wireless devices and the frequencies that they run on) that incorporates protection against non-thermal effects of radio frequency on living species, we will continue to be at loggerheads with regard to making any meaningful guidelines that will protect us.

In USA FCC, the specific absorption rate limits of exposure to radio frequency in the range of 100 kHz to 6 GHz for the general public and for those who get exposed by virtue of their occupation have been laid down (Appendix XII).

Far-field exposure from cell towers has steadily increased as more and more towers are being erected in neighbourhoods. Consequent to the revision of EMF radiation norms by the Department of Telecommunications (DoT), the Indian standards are now ten times more stringent than many countries like USA, Canada, Japan, and Australia. Most countries in the world follow ICNIRP guidelines. A number of countries have specified their own radiation norms for exposure limits from mobile towers, keeping in mind the environmental and physiological factors (Appendix XIII).

Graphic warning labels on cigarette packs became a statutory requirement after decades of persuasion by the scientific community. A warning on the back of the body of mobile phone, such as 'Hold this device at least an inch away from your body', should be displayed more prominently rather than be concealed in the smallest font in the Setting option of the phone would allow the user to make an informed choice.

I am not an engineer, but it is conceivable to build a technology into the design of the phone, such as a warning beep, that would make it impossible to operate unless held at least an inch from the body.

Several countries have issued initiatives about the safe use of mobile phones and wireless devices at home and at the workplace. However, an advisory is not a law.

Ironically, the initiative to protect wildlife and plants from radio frequency radiation is not even a matter of debate. Bird migration is a phenomenon in nature since time immemorial. There is clearly a significant decline in population of migratory birds, bees, bats, and many other forms of wildlife.

In the meanwhile, prudent avoidance is the key. The following steps are recommended to minimize harm in case of near-field exposure:

1. When not in use, stow the mobile phone away from the body in a purse or handbag. If you have a pacemaker, never keep the phone in the breast pocket. It can stop the pacemaker or cause the pacemaker to discharge pulses either at a fixed rate or irregularly. Young males should not keep the phone in the pants pocket. If testes are exposed, there is risk of infertility.
2. Do not use the phone next to your ear.
3. Use a speakerphone or an air tube. Air tube is preferred over earphones.
4. Keep the phone charged. Do not use it if battery is running out. At full charge, the phone minimizes use of power.
5. Avoid using mobile phones where signal is poor. When signal is poor, the phone tends to maximize its power in order to connect with a cell tower.
6. When travelling in a car, train, or an elevator, use the phone only for emergencies. In a closed metal box, the EM waves are amplified.
7. Keep duration of the call to the minimum possible. Use texting as much as possible. While texting, phones use much less energy.
8. Never sleep with a phone under the pillow
9. Young children, pregnant women, and the elderly are at greater risk.

10. Do not use the phone while driving.
11. Cordless phones have similar disadvantages as mobile phones. Use the landline as much as possible. Using the Call Forward feature, cell phone calls can be forwarded to the landline.
12. There is no evidence that radiation shields protect against radiation. Indeed, they may interfere with the functioning of mobile phone, causing it to use more power.
13. Use corded connections for printers, mouse, gaming, keyboard, etc.
14. If the device is in the hands of a child, turn off Wi-Fi and Bluetooth, and put the device in airplane mode.
15. Do not hold iPads, laptops, tablets on your lap.
16. Schools should have a wired LAN access for all devices.
17. Turn off Wi-Fi router before you sleep. Persuade your neighbours to do the same.
18. Consider grounding. The principle of grounding is to reconnect with the earth. Ever since shoes were discovered, we have lost contact with bare earth. Earth is a profound source of electrons. As we absorb EMFs, we become positively charged, and our body voltage rises. This can be measured with body voltage metres. In perfect health, our body should be electrically neutral or as close to neutral as possible. Re-establishing contact with earth by simply walking barefoot allows our bodies to absorb electrons from earth and thus become electrically neutral. This can minimize free radical–induced oxidative stress significantly. There are, indeed, devices designed, such as grounding bed sheets and mats, which can be used in high-rise buildings. Such devices can be earthed by connecting them to grounding contact opening of a wall outlet.

Appendix I

Frequency, wavelength, and energy chart of electromagnetic spectrum

Region	Frequency (Hz)	Wavelength (m)	Energy (eV)
Radio waves	$<10^9$	>0.3	$<7 \times 10^{-7}$
Microwaves	$10^9–3 \times 10^{11}$	$0.001–0.3$	$7 \times 10^{-7}–2 \times 10^{-4}$
Infrared	$3 \times 10^{11}–3.9 \times 10^{14}$	$7.6 \times 10^{-7}–0.001$	$2 \times 10^{-4}–0.3$
Visible	$3.9 \times 10^{14}–7.9 \times 10^{14}$	$3.8 \times 10^{-7}–7.7 \times 10^{-7}$	$0.3–0.5$
UV	$7.9 \times 10^{14}–3.4 \times 10^{16}$	$8 \times 10^{-9}–3.8 \times 10^{-7}$	$0.5–20$
X-rays	$3.4 \times 10^{16}–5 \times 10^{19}$	$6 \times 10^{-12}–8 \times 10^{-9}$	$20–3 \times 10^4$
Gamma rays	$>5 \times 10^{19}$	$<6 \times 10^{-12}$	$>3 \times 10^4$

Appendix II

Measurements and Conversion Tables

Frequency

1GHz = 1,000 MHz
1 MHz = 1,000 kHz
1 kHz = 1,000 Hz

Magnetic fields

1 ampere/metre	=	1000 milliampere/metre
1.254 microtesla (μT)		
		0.001254 millitesla (mT)
		0.000001 tesla (T)
1 gauss	=	1000 milli gauss (mG)
0.0001 tesla (T)		
0.1 millitesla (mT)		
100 microtesla (μT)		
1 tesla	=	10,000 gauss
		1000 millitesla (mT)
		1000,000 microtesla (μT)
		1000,000,000 nanotesla (nT)

As most environmental EMFs produce very tiny magnetic fields, we need smaller units to quantify them. Therefore, units of practical use are microtesla (μT) or milligauss (mG).

To convert microtesla (μT) to milligauss (mG):
1 μT= 10 mG
0.1 μT = 1 mG

Gauss and *tesla* describe magnetic flux density. Magnetic flux density is the amount of magnetic flux in an area taken perpendicular to its direction. It is denoted as *B*. Magnetic field strength, defined by its magnitude in a given direction, is denoted as *H*.

$B = \mu H$, where μ is magnetic permeability or magnetizability

The average intensity of earth's geomagnetic field is 500 mG or 0.5 G.

Electric fields

AC electric field is measured as volt/metre and watt/metre^2.
watt/metre^2 = (Volts/m)^2 ÷ 377
1 watt (W) = 1,000 milliwatts (mW)
1 milliwatt (mW) = 1,000 microwatts (μW) or
1 watt (W) = 1,000,000 microwatts (μW)

Power density (Pd) defines the quantum of energy in an electromagnetic wave and is measured as follows: *Pd* = *E* (Volts/m) x *H* (Amp/m), where *E* is the electric field strength and *H* is magnetic field strength.

The average intensity of the static electric field of the earth is 130 V/m.

Radio frequency fields

Volt/m is the most practical unit for measurement of RF fields and AC electric fields at non-thermal levels. In case of low-intensity fields, watt (W), milliwatt (mW), and microwatt (µW) per unit area are used.

- microWatts per square metre (µW/m2)
- microWatts per square centimetre (µW/cm2)
- milliWatts per square metre (mWm^2)
- volts per metre (V/m).

mW means 'milliwatts', and *µW* means 'microwatts'.
$1\ W/m^2 = 1{,}000\ mW/m^2 = 1{,}000{,}000\ \mu W/m^2$
$1\ \mu W/cm^2 = 10{,}000\ \mu W/m^2 = 10\ mW/m^2$

Appendix III

Exposure limits for the general public for 50 Hz power frequency fields in European Union Recommendation 1999/519/EC (2011)

Country	Electric field strength (V/m)	Magnetic flux density (μT)
Recommendation	5,000	100
Austria	5,000	100
Czech Rep	5,000	100
Estonia	5,000	100
France	5,000	100
Hungary	5,000	100
Ireland	5,000	100
Luxembourg	5,000	100
Malta	5,000	100
Romania	5,000	100
Slovakia	5,000	100
Britain	5,000	100
Germany	5,000	100

Source: National Institute for Public Health and the Environment, Ministry of Health, Welfare and Sport, Netherlands.

Appendix IV

Exposure limits for the occupational exposure for 50 Hz power frequency fields in European Union Directive 2004/40/EC (2011)

Country	Electric field strength (V/m)	Magnetic flux density (μT)
Recommendation	10,000	500
Austria	10,000	500
Czech Rep	10,000	500
Denmark	10,000	500
France	10,000	500
Hungary	10,000	500
Italy	10,000	500
Latvia	10,000	500
Malta	10,000	500
Romania	10,000	500
Slovakia	10,000	500
Britain	10,000	500
Germany	21,320	1,358

Luxembourg	5,000	100
USA	25,000	1,000

Source: National Institute for Public Health and the Environment, Ministry of Health, Welfare and Sport, Netherlands.

Appendix V

Exposure limits for the general public for MW RF EMFs for 900MHz GSM in European Union Recommendation 1999/519/ EC (2011)

Country	Electric field strength (V/m)	Magnetic flux density (μT)	Equivalent power density (W/m^2)
Recommendation	41	0.14	4.5
Austria	41	0.14	4.5
Czech Rep	41	0.14	4.5
Finland	41	0.14	4.5
France	41	0.14	4.5
Hungary	41	0.14	4.5
Ireland	41	0.14	4.5
Malta	41	0.14	4.5
Romania	41	0.14	4.5
Slovakia	41	0.14	4.5
Britain	41	0.14	4.5
Germany	41	0.14	4.5

Luxembourg	41	0.14	4.5
Portugal	41	0.14	4.5

Source: National Institute for Public Health and the Environment, Ministry of Health, Welfare and Sport, Netherlands.

Appendix VI

Exposure limits for the general public for MW RF EMFs for 1800MHz GSM in European Union Recommendation 1999/519/EC (2011)

Country	Electric field strength (V/m)	Magnetic flux density (μT)	Equivalent power density (W/m²)
Recommendation	58	0.20	9
Austria	58	0.20	9
Czech Rep	58	0.20	9
Finland	58	0.20	9
France	58	0.20	9
Hungary	58	0.20	9
Ireland	58	0.20	9
Malta	58	0.20	9
Romania	58	0.20	9
Slovakia	58	0.20	9
Britain	58	0.20	9
Germany	58	0.20	9
Luxembourg	58	0.20	9
Portugal	58	0.20	9

Italy	06	0.02	0.1
Slovenia	18	0.06	0.9

Source: National Institute for Public Health and the Environment, Ministry of Health, Welfare and Sport, Netherlands.

Appendix VII

Exposure limits for the general public for MW RF EMFs for 2100MHz UTMS in European Union Recommendation 1999/519/EC (2011)

Country	Electric field strength (V/m)	Magnetic flux density (μT)	Equivalent power density (W/m^2)
Recommendation	61	0.20	10
Austria	61	0.20	10
Czech Rep	61	0.20	10
Finland	61	0.20	10
France	61	0.20	10
Hungary	61	0.20	10
Ireland	61	0.20	10
Malta	61	0.20	10
Romania	61	0.20	10
Slovakia	61	0.20	10
Britain	61	0.20	10
Germany	61	0.20	10
Luxembourg	61	0.20	10
Portugal	61	0.20	10

Italy	06	0.02	0.1
Slovenia	19	0.06	01

Source: National Institute for Public Health and the Environment, Ministry of Health, Welfare and Sport, Netherlands.

Appendix VIII

Exposure limits for the occupational exposure for MW RF EMFs for 900MHz GSM in European Union Directive 2004/40/EC (2011)

Country	Electric field strength (V/m)	Magnetic flux density (μT)	Equivalent power density (W/m^2)
Directive	90	0.30	22.5
Austria	90	0.30	22.5
Czech Rep	90	0.30	22.5
Denmark	90	0.30	22.5
Finland	90	0.30	22.5
France	90	0.30	22.5
Hungary	90	0.30	22.5
Italy	90	0.30	22.5
Latvia	90	0.30	22.5
Lithuania	90	0.30	22.5
Malta	90	0.30	22.5
Switzerland	90	0.30	22.5
Britain	90	0.31	22.5
Germany	92	0.30	22.5

Luxembourg	41	0.14	4.5
Sweden	60	—	10

Source: National Institute for Public Health and the Environment, Ministry of Health, Welfare and Sport, Netherlands.

Appendix IX

Exposure limits for the occupational exposure for MW RF EMFs for 1800MHz GSM in European Union Directive 2004/40/EC (2011)

Country	Electric field strength (V/m)	Magnetic flux density (μT)	Equivalent power density (W/m^2)
Directive	127	0.42	45
Austria	127	0.42	45
Czech Rep	127	0.42	45
Denmark	127	0.42	45
Finland	127	0.42	45
France	127	0.42	45
Hungary	127	0.42	45
Italy	127	0.42	45
Latvia	127	0.42	45
Lithuania	127	0.42	45
Malta	127	0.42	45
Switzerland	127	0.42	45
Britain	127	0.42	45
Germany	130	0.43	45

Luxembourg	58	0.20	9
Sweden	60	—	10

Source: National Institute for Public Health and the Environment, Ministry of Health, Welfare and Sport, Netherlands.

Appendix X

Exposure limits for the occupational exposure for MW RF EMFs for 2100MHz UTMS in European Union Directive 2004/40/EC (2011)

Country	Electric field strength (V/m)	Magnetic flux density (μT)	Equivalent power density (W/m^2)
DIRECTIVE	137	0.45	50
Austria	137	0.45	50
Czech Rep	137	0.45	50
Denmark	137	0.45	50
Finland	137	0.45	50
France	137	0.45	50
Hungary	137	0.45	50
Italy	127	0.42	45
Latvia	137	0.45	50
Lithuania	137	0.45	50
Malta	137	0.45	50
Romania	137	0.45	50
Slovakia	137	0.45	50
Switzerland	137	0.45	50
Britain	137	0.45	50

Germany	137	0.46	50
Luxembourg	61	0.20	10

Source: National Institute for Public Health and the Environment, Ministry of Health, Welfare and Sport, Netherlands.

Appendix XI

FCC limits for maximum permissible exposure (MPE) 1996 occupational exposure

Frequency Range (MHz)	Electric Field (V/m)	Magnetic Field (A/m)	Power Density (mW/cm^2)	Average Time (minutes)
0.3–3.0	614	1.63	100	6
3.0–30	1842/f	4.89/f	900/f	6
30–300	61.4	0.163	1.0	6
300–1500	—	—	f/300	6
1500–100,000	—	—	5	6

Source: FCC OET Bulletin 56 Fourth Edition.

General public

Frequency Range (MHz)	Electric Field (V/m)	Magnetic Field (A/m)	Power Density (mW/cm^2)	Average Time (minutes)
0.3–1.34	614	1.63	100	30
1.34–30	842/f	2.19/f	180/f	30
30–300	27.5	0.073	0.2	30

300–1,500	—	—	f/1500	30
1,500–100,000	—	—	5	30

Source: FCC OET Bulletin 56 Fourth Edition.
NB: 1 Ampere/metre = 1.254 μT.

Appendix XII

FCC SAR limits for localized (partial body) exposure in 100 kHz–6 GHz

Occupational	General population
<0.4 W/kg whole body	<0.08 W/kg whole body
<8 W/kg partial body	<1.6 W/kg partial body

Source: FCC OET Bulletin 56 Fourth Edition.

ICNIRP guidelines on exposure 1998 with reference to power lines 60 Hz

	Occupational	General population
Electric field	8.3 kV/m	4.2 kV/m
Magnetic field	4200 mG	833 mG

Source: ICNIRP Guidelines, 1998.

Appendix XIII

Table 1 International EMF radiation norms for mobile towers international exposure limits for EMF in W/m^2 (1,800 MHz)

USA, Canada, and Japan	12
ICNIRP and EU recommendation 1998	9.2
Exposure limit in Australia	9
Exposure limit in Belgium	2.4
Exposure limit in Italy, Israel	1.0
Exposure limit in Auckland, New Zealand	0.5
Exposure limit in Luxembourg	0.45
Exposure limit in China	0.4
Exposure limit in Russia, Bulgaria	0.2
Exposure limit in Poland, Paris, Hungary	0.1
Exposure limit in Italy in sensitive areas	0.1
Exposure limit in Switzerland	0.095
ECOLOG 1998 (Germany)	0.09
Exposure limit in Austria	0.001

Source: TRAI, 2014

Appendix XIV

Magnetic fields from home appliances (mG)

Vacuum cleaner	700 mG
Hair dryer	700 mG
Mixers	600 mG
Shaver	600 mG
Microwave oven	300 mG
Electric cooking range	200 mG
Food processor	130 mG
Washing machine	100 mG
Blender	100 mG
CFL lights	100 mG
Refrigerator	40 mG
Electric blanket	40 mG
Window AC	20 mG
Iron	20 mG
Digital clock	8 mG

Magnetic fields from some occupations (mG)

Use of power saw	1,000 mG
Substation operators	7.2 mG

Electricians	5.4 mG
TV repairers	4.3 mG

Source: EMF RAPID 2002, National Institute of Environmental Health Sciences, USA.

Exposure levels from various sources (microwatt/sq.cm)

Source	Exposure Level
Cosmic background radiation	.000000000000000008
Quiet sun	.0000000001
Satellite	.0000001
300 feet from a cell tower	5.0
FCC safety standard	1000.0

Glossary

AC	alternating current as opposed to direct current
APC	adaptive power control, a technology in newer mobile phones to minimize power consumption
Agent Orange	a synthetic weed killer herbicide manufactured by Monsanto and Dow Chemicals and used by the United States Army during Operation Ranch Hand in the Vietnam War to deprive guerrillas of hideouts and food
ALARA	as low as reasonably achievable, a safety principle and a statutory requirement to minimize radiation dose
amino acid	an organic molecule comprising a carboxyl (—COOH) and an amino group (—NH_2), which is a primary constituent of proteins
AMPS	Advanced Mobile Phone System, earliest analogue system, now discontinued
aneuploidy	the presence of abnormal number of chromosomes in a cell other than the normal number (46) as a consequence of mutation
anion	a state of an atom in which electrons outnumber protons, causing it to be negatively charged
apoptosis	programmed death of cells
ANSI	American National Standards Institute

antioxidant	a molecule to counteract oxidation; oxidation is a biochemical process that gives rise to production of free radicals as a chain reaction and has both useful and damaging effects (melatonin, vitamin C, and trolox are examples of antioxidants)
ATC	air traffic control
AT&T	American Telephone and Telegraph, founded in 1885
Big Bang	most widely accepted theory of the origin of the universe
BBB	blood–brain barrier, a natural arrangement in the human and animal brain to protect it from invasion of toxins circulating in the general circulation of blood
BMI	body mass index
Biofield	an energy field surrounding the body of all living species and emanating from electromagnetic activity of the body
biological clock	a system of genes and proteins that work in a clockwork manner to drive the circadian rhythm
brick phone	DynaTAC, the first mobile phone
cAMP	cyclic adenosine monophosphate, a messenger molecule used in carrying effects of other molecules into cells which cannot get through cell walls directly, such as hormones
carbon footprint	the emission of greenhouse gases individually and collectively by people, events, industries, etc.; it's measured as equivalent tons of CO_2
case control study	study of 'cases' versus 'controls' by retrospective analysis of the characteristics of exposure to the causative agent
cation	a state of an atom in which the protons outnumber the electrons, causing it to be positively charged

CCD	colony collapse disorder, an epidemic observed in honeybees, forcing them to desert their hives as a result of exposure to radio frequency radiation from mobile phone masts
CDMA	code division multiple access, a protocol used in digital wireless communication in order to carry multiple signals in one channel
CEET	Centre for Energy Efficient Telecommunication (Australia)
CEFALO	a case control study of risk of brain tumours in children and adolescents due to mobile phone use, conducted in Switzerland and Scandinavian countries between 2004 and 2008
Cell Lines	assembly of cells cloned from one cell obtained by biopsy of tissue; they can be stored indefinitely in liquid nitrogen and thus provide abundant supply for genetic studies
CERENAT	a French case control study of association between brain tumours and the use of mobile phones in adults, conducted between 2004 and 2006
CFL	compact florescent light
charge	a property of matter to experience a force when placed in an electromagnetic field
chromosome	twenty-two pairs of thread-like somatic strands of DNA plus two strands for germ cells (XX for female and XY for male) for a total of 46.
chronobiology	the field of biology that examines periodic events in the life of living beings, such as circadian rhythms
circadian rhythm	a 24-hour cycle in the life of all living beings that governs sleeping, waking, and feeding patterns among other functions

clusterin	a protein, encoded by CLU gene on chromosome 8 and entrusted the task to scavenge the cellular debris after cells die a natural death
CMB	cosmic microwave background, radiation that exists today in the background, left over from the union of charged subatomic particles to form electrically neutral hydrogen atoms for the first time about 380,000 years after Big Bang
cohort study	longitudinal prospective study of a group (cohort) of people with similar demographic characteristics and are exposed to a certain risk factor to establish correlation if and when they develop disease
COMEST	Commission on Ethics of Scientific Knowledge and Technology (set up by United Nations)
comet assay	a laboratory technique to demonstrate DNA damage; it's also known as single-cell gel electrophoresis (SCGE)
cross-sectional study	analysis of data from a given population to establish correlation between risk factor and disease at a given point in time
current	The flow of charge through a material such as a wire
CSSAD	Committee for Scientific Survey of Air Defence (UK)
CTIA	Cellular Telephone Industry Association
DC	direct current, as opposed to alternating current
DCS	digital communication system
DECT	Digital Enhanced Cordless Telecommunication
digital dementia	cognitive decline due to overexposure to wireless technology
DNA	deoxyribonucleic acid
drosophila	fruit fly, a model organism for genetic studies

DynaTAC	Dynamic Adaptive Total Area Coverage (first mobile phone, 1973)
ECG	electrocardiogram
EEG	electroencephalogram
EHS	electrohypersensitivity
ELF	extremely low frequency (3–300 Hz)
EMF	electromagnetic Field
EMR	electromagnetic radiation
EMW	electromagnetic wave
epidemiology	study of causes, transmission, and effects of disease in a given population
Faraday's law	a law to conjecture how an electromotive force is induced in an electric circuit when exposed to a magnetic field
Far field	the field in which distance from the antenna exceeds the wavelength of the radiated EMF
Fenton reaction	an iron-mediated catalytic process that generates reactive oxygen species (free radicals)
FCC	Federal Communications Commission (USA)
FDA	Food and Drug Administration (USA)
free radical	an atom, ion, or molecule having an unpaired electron and therefore highly reactive
frequency	number of cycles per second
GBM	glioblastoma multiforme
GE	General Elelctric
gene	a region on chromosomes responsible for coding for proteins
GMF	geomagnetic field
GMOs	genetically modified organisms
GSH-Px	glutathione peroxidase
GSM	Global System for Mobile Communication
HAARP	High Frequency Active Auroral Research Program

Hill criteria	set of conditions to establish correlation between stated cause and its effect in medical research
HSP	heat shock proteins, a class of proteins elaborated by living cells in response to stress
IARC	International Association on Cancer Research (under WHO)
ICD-10	International Classification of Diseases (2010)
ICNIRP	International Commission on Nonionizing Radiation Protection
ICT	information and communication technology
IEEE	Institute of Electrical and Electronic Engineers
Interphone study	case control study conducted from 2000 to 2006 by IARC to establish link between mobile phone use and brain tumours
in vitro	a biological experiment conducted outside the living body in the laboratory
in vivo	a biological experiment conducted inside the living body
ionizing radiation	high-energy X-ray and gamma ray electromagnetic radiation that can ionize living tissue by cleaving electrons from its atoms
IoT	Internet of things
JAMA	*Journal of American Medical Association*
LCD	liquid crystal display
LED	light-emitting diode
Living Planet Index	An index issued by ZSL (see below) that keeps track of over 10,000 species of animals worldwide with regard to environmental impact on species
LTE	Long Term Evolution with reference to 4G
M2M	machine-to-machine communication
Luftwaffe	the German air force

magnetoreception	the ability of living species to perceive geomagnetic field
MCG	magnetocardiography
MCS	multiple chemical sensitivity
MEG	magnetoencephalography
MK-ULTRA	an illegal mind control programme used by the CIA (USA) in the 1950s to develop drugs for use in future interrogations
melanopsin	a light pigment found in the eyes of mammals, sensitive to blue light; it's not involved in vision but is part of the system responsible for the maintenance of the circadian rhythm
melatonin	a powerful natural antioxidant hormone released by the pineal gland that maintains the sleep–wake cycle and helps in cellular repair
micronucleus	formed when during cell division, some bits of chromosomes lie outside the nucleus instead of being retained within the nuclei of daughter cells; may be a sign of developing cancer
microwave oven	an appliance used to cook food with electromagnetic radiation in the microwave range
modem	modulator and demodulator
MRI	magnetic resonance imaging, a radiological investigative technique
mutation	a process that changes nucleotide sequence of genes, causing a mutated gene, which behaves abnormally
MW	microwave, nonionizing radiation between 300MHz to 300 GHz frequency range
NDRC	National Defense Research Committee (USA)
near field	the field wherein distance from the antenna is less than the wavelength of EMF radiation

NFC	near-field communication, a newer feature of mobile phones
NIH	National Institute of Health (USA)
NMT	Nordic Mobile Telephony, first analogue 1G but automated system of cellular telephony launched in 1981 in Sweden and Denmark
nonionizing radiation	electromagnetic radiation that does not contain enough energy to cleave an electron from an atom
NRPB	National Radiological Protection Board, a public authority (UK)
NTP	National Toxicology Program (USA)
Nuremberg Code	a set of ten ethical and statutory principles for human experimentation brought into effect at the Nuremberg trials held by the USA in Germany at the end of Second World War
OFDM	orthogonal frequency division multiplexing, a protocol used in multicarrier digital wireless communication in order to carry multiple signals in several parallel streams
oxidation	loss of an electron by an atom, ion, or molecule
oxidative stress	cellular stress caused when generation of free radicals fails to be counteracted by antioxidants
Paris Agreement	an agreement signed by 195 nations on 12 December 2015, in Paris under the auspices of the United Nations to curb global warming
photoperiodism	physiological response of living species to the length of day or night
pineal gland	a gland in the brain of vertebrates that produces the hormone melatonin
power density	quantum of energy per unit area in an electromagnetic wave (Pd (watts/square metre) = E (volts/m) × H (Amp/m)); E is electric field strength, and H is the magnetic field strength

precautionary principle	an advisory by policy makers to the public to take precautions against a suspected agent or product while scientific consensus is pending on its being harmful or otherwise
protein folding	an array of shapes and designs in which amino acids link up to build proteins, each with a specific function
psSAR	peak spatial SAR averaged over 1 g or 10 g of tissue
radar	radioactive detection and ranging, a microwave-based technology in which microwave beams are fired and their reflection from metallic objects (such as ships and aircraft or vehicles) is used to calculate their location, angle, altitude, and direction
RAF	Royal Air Force (UK)
RCCs	Radio Common Carriers, a conglomerate of companies smaller than AT&T in USA
reduction	gain of an electron by an atom, ion, or molecule
redox reaction	reduction–oxidation reaction
REFLEX Study	A multicentre study conducted between 2000 and 2004 in Germany to see if ELF EMFs and RF EMFs had genotoxic effects on cells
RF	radio frequency, nonionizing radiation (300 Hz to 300 GHz) useful for radio communication
RFID	radio frequency identification, a newer technology to embed virtually any object and use it to transmit data to the Internet of things via wireless cloud
RMS *Titanic*	a luxury British ship that sank in the Atlantic Ocean on 14 April 1912
RNA	ribonucleic Acid, a messenger molecule that carries instructions from DNA to ribosome to build proteins

RNS	reactive nitrogen species, a type of free radical
ROS	reactive oxygen species, a type of free radical
SAR	specific absorption rate is the rate at which the body absorbs radio frequency energy when exposed to a radio frequency electromagnetic field; it is measured in microwatts per square centimetre, adopted for frequencies above 100 KHz
SAM	specific anthropomorphic mannequin, used to calculate SAR
Schumann resonances	natural ELF EMFs at 7.83 Hz due to electric tension between earth and the ionosphere
SCN	suprachiasmatic nucleus at the base of hypothalamus that controls the mechanisms involved in the circadian rhythm
smart antenna	multiple-input-multiple-output (MIMO) array of antennas capable of several simultaneous functions
smart metre	a microwave-based communication system to monitor the consumption of electricity and water in a dwelling and relay it to the provider company
solar radiation	very high energy radiation from the sun, such as solar winds
SOD	superoxide dismutase
stem cells	parent undifferentiated cells found in all multicellular organisms that can divide to produce more stem cells but can also differentiate into cells of other kinds depending on their location
strand breaks	physical breaks in DNA strands
STUK	Radiation Safety Authority in Finland
TACS	Total Access Communication System, a type of AMPS, now discontinued

TEM cells	transverse electromagnetic cells, used in research on mice as chambers for controlled exposure to RF
TNF a	tumour necrosis factor alpha
UMTS	Universal Mobile Telecommunications System
UNMC	University of Nebraska Medical Centre
USEPA	United States Environmental Protection Agency
UV radiation	ultraviolet radiation, a natural ionizing radiation
valency	affinity of an atom to bind with a neighbouring atom
VGCC	voltage-gated calcium channel
WAN	wide area network
wavelength	distance between two crests of one wave
wbSAR	whole-body SAR, total EMF absorbed by a body divided by its mass
WHO	World Health Organization
WiMax	worldwide interoperability for microwave access, a newer technology to provide wireless access over very large areas
Wistar rat	cute-looking, red-eyed, white-furred rats specially bred for research first in the late nineteenth century in Wistar Institute of the University of Philadelphia, Pennsylvania
WSN	wireless sensor network
WTR	wireless technology research, a $25 million CTIA-financed research programme of dubious repute supposedly conducted in 1994 and headed by George Carlo
XO	xanthine oxidase, an antioxidant
zettabyte	10^{21} bytes

ZSL	Zoological Society of London, founded in 1826, devotes itself to the conservation of animals worldwide, conducts research along with World Wide Fund for Nature (WWF), and issues periodic Living Planet Index

Endnotes

Chapter 5: The Sar Hoax

1 O. P. Gandhi et al., 'Exposure Limits: The Underestimation of Absorbed Cell Phone Radiation, especially in Children', *Electromagnetic Biology and Medicine*, 31/1 (2012), 34–51.

2 Allan H. Frey, 'Human Auditory System Response to Modulated Electromagnetic Energy', *Journal of Applied Physiology*, 17/ 4 (1962), 689–692.

Chapter 6: Conundrum of Research

1 N. Wertheimer, E. Leeper, 'Electrical Wiring Configurations and Childhood Cancer, *American Journal of Epidemiology*, 109/3 (March 1979), 273–84.

2 S. Milham, E. M. Ossiander, 'Historical Evidence That Residential Electrification Caused the Emergence of the Childhood Leukemia Peak', *Medical Hypotheses*, 56/3 (2001), 290–295.

3 S. Milham, 'Historical Evidence that Electrification Caused the 20th Century Epidemic of 'Diseases of Civilization', *Medical Hypotheses*, 74 (2010), 337–345.

4 M. Feychting, F. Jonsson, N. L. Pedersen, A. Ahlbom, 'Occupational Magnetic Field Exposure and Neurodegenerative Disease', *Epidemiology*, 14/4 (July 2003), 413–9, discussion 427–8.

5 M. Kundi, K. Mild, L. Hardell, M. O. Mattsson, 'Mobile Telephones and Cancer—A Review of Epidemiological Evidence', *Journal of Toxicology and Environmental Health Part B Critical Reviews*, 7/5 (September–October 2004), 351–84.

[6] J. E. Muscat, M. G. Malkin, S. Thompson, R. E. Shore, S. D. Stellman, D. McRee, A. I. Neugut, E. L. Wynder, 'Handheld Cellular Telephone Use and Risk of Brain Cancer', *Journal of the American Medical Association*, 284/23 (December 2000), 3001–7.

[7] P. D. Inskip PD, R. E. Tarone, E. E. Hatch, T. C. Wilcosky, W. R. Shapiro, R. G. Selker, H. A. Fine, P. M. Black, J. S. Loeffler, M. S. Linet, 'Cellular-Telephone Use and Brain Tumors', *New England Journal of Medicine*, 344/2 (January 2001), 79–86.

[8] J. Siemiatycki, L. Richardson, K. Straif, B. Latreille, R. Lakhani, S. Campbell, M. C. Rousseau, P. Boffetta, 'Listing Occupational Carcinogens,' *Environmental Health Perspectives*, 112/15 (November 2004), 1447–59.

[9] Interphone Study Group, 'Brain Tumour Risk in Relation to Mobile Telephone Use: Results of the Interphone International Case-Control Study', *International Journal of Epidemiology*, 39/3 (June 2010), 675–94.

[10] S. Sadetzki, A. Chetrit, A. Jarus-Hakak, E. Cardis, Y. Deutch, S. Duvdevani, A. Zultan, I. Novikov, L. Freedman, 'Cellular Phone Use and Risk of Benign and Malignant Parotid Gland Tumors—A Nationwide Case-Control Study', *American Journal of Epidemiology*, 167/4 (15 February 2008), 457–67.

[11] D. Aydin, M. Feychting, J. Schüz, T. Tynes, T. V. Andersen, L. S. Schmidt, A. H. Poulsen, C. Johansen, M. Prochazka, B. Lannering, L. Klæboe, T. Eggen, D. Jenni, M. Grotzer, N. Von der Weid, C. E. Kuehni, M. Röösli, 'Mobile Phone Use and Brain Tumors in Children and Adolescents: A Multicenter Case-Control Study', *Journal of the National Cancer Institute*, 103/16 (17 August 2011), 1264–76.

[12] F. Söderqvist, M. Carlberg, K. Hansson Mild, L. Hardell, 'Childhood Brain Tumour Risk and Its Association with Wireless Phones: A Commentary', *Environmental Health*, 10 (19 December 2011), 106.

[13] D. Pettersson, T. Mathiesen, M. Prochazka, T. Bergenheim, R. Florentzson, H. Harder, G. Nyberg, P. Siesjö, M. Feychting, 'Long-Term Mobile Phone Use and Acoustic Neuroma Risk', *Epidemiology*, 25/2 (March 2014), 233–41.

[14] M. Feychting, 'Invited Commentary: Extremely Low-Frequency Magnetic Fields and Breast Cancer—Now It Is Enough!', *American Journal of Epidemiology*, 178/7 (1 October 2013), 1046–50.

[15] Fang et al., 'Misconduct Accounts for the Majority of Retracted Scientific Publications', *Proceedings of the National Academy of Sciences of the United States of America*, 109/42 (2012), 17028–17033.

[16] M. P. Little, 'Cancer and Non-Cancer Effects in Japanese Atomic Bomb Survivors', *Journal of Radiological Protection*, 29/2A (June 2009), A43–59.

[17] L. Hardell, M. Carlberg, K. Hansson Mild, 'Pooled Analysis of Two Case-Control Studies on Use of Cellular and Cordless Telephones and the Risk for Malignant Brain Tumours Diagnosed in 1997–2003', *International Archives of Occupational and Environmental Health*, 79/8 (September 2006), 630–9.

[18] L. Hardell, K. H. Mild, M. Carlberg, F. Söderqvist, 'Tumour Risk Associated with Use of Cellular Telephones or Cordless Desktop Telephones', *World Journal of Surgical Oncology*, 4 (11 October 2006), 74.

[19] L. Hardell, M. Carlberg, F. Söderqvist, K. H. Mild, L. L. Morgan, 'Long-Term Use of Cellular Phones and Brain Tumours: Increased Risk Associated with Use for > or =10 Years', *Occupational and Environmental Medicine*, 64/9 (September 2007), 626–32.

[20] L. Hardell, M. Carlberg, K. Hansson Mild, 'Epidemiological Evidence for an Association between Use of Wireless Phones and Tumor Diseases', *Pathophysiology*, 16/2–3 (August 2009), 113–22.

[21] V. G. Khurana, C. Teo, M. Kundi, L. Hardell, M. Carlberg, 'Cell Phones and Brain Tumors: A Review Including the Long-Term Epidemiologic Data', *Surgical Neurology*, 72/3 (September 2009), 205–14; discussion 214–5.

[22] V. G. Khurana, L. Hardell, J. Everaert, A. Bortkiewicz, M. Carlberg, M. Ahonen, 'Epidemiological Evidence for a Health Risk from Mobile Phone Base Stations', *International Journal of Occupational and Environmental Health*,16(3) (Jul-Sep 2010), 263-7.

[23] L. Hardell, M. Carlberg, 'Mobile Phone and Cordless Phone Use and the Risk for Glioma—Analysis of Pooled Case-Control Studies in

Sweden, 1997–2003 and 2007–2009', *Pathophysiology*, 22/1 (March 2015), 1–13.

[24] M. Carlberg, L. Hardell, 'Pooled Analysis of Swedish Case-Control Studies during 1997–2003 and 2007–2009 on Meningioma Risk Associated with the Use of Mobile and Cordless Phones', *Oncolgy Reports*, 33/6 (June 2015), 3093–8.

[25] R. Wolf, D. Wolf, 'Increased Incidence of Cancer Near a Cell-Phone Transmitter Station', *International Journal of Cancer Prevention*, 1/2 (2004), 123–128.

[26] L. Hardell, M. J. Walker, B. Walhjalt, L. S. Friedman, E. D. Richter, 'Secret Ties to Industry and Conflicting Interests in Cancer Research', *American Journal of Industrial Medicine*, 50/3 (March 2007), 227–33.

[27] L. Hardell, M. Carlberg, 'Using the Hill Viewpoints from 1965 for Evaluating Strengths of Evidence of the Risk for Brain Tumors Associated with Use of Mobile and Cordless Phones', *Reviews on Environmental Health*, 28/2–3 (2013), 97–106.

[28] International Agency for Research on Cancer-(IARC), 'Non-Ionizing Radiation, Part II, Radiofrequency Electromagnetic Fields (RF-EMF)', *Monograph*, 102 (2011).

[29] A. J. Swerdlow, M. Feychting, A. C. Green, L. K. Leeka Kheifets, D. A. Savitz, 'Mobile Phones, Brain Tumors, and the Interphone Study: Where Are We Now?', *Environmental Health Perspectives*, 119/11 (November 2011), 1534–8.

[30] G. Coureau, G. Bouvier, P. Lebailly, P. Fabbro-Peray, A. Gruber, K. Leffondre, J. S. Guillamo, H. Loiseau, S. Mathoulin-Pélissier, R. Salamon, I Baldi, 'Mobile Phone Use and Brain Tumours in the CERENAT Case-Control Study', *Occupational and Environmental Medicine*, 71/7 (July 2014), 514–22.

[31] L. L. Morgan, A. B. Miller, A. Sasco, D. L. Davis, 'Mobile Phone Radiation Causes Brain Tumors and Should Be Classified as a Probable Human Carcinogen (2A) (Review)', *International Journal of Oncology*, 46/5 (May 2015), 1865–71.

[32] G. Zada et al., 'Incidence Trends in the Anatomic Location of Primary Malignant Brain Tumors in the United States: 1992–2006', *World Neurosurgery*, 77 (2012), 518–524.

33 M. Dobes et al., 'Increasing Incidence of Glioblastoma Multiforme and Meningioma, and Decreasing Incidence of Schwannoma (2000–2008): Findings of Multicenter Australian Study', *Surgical Neurology International*, 2 (2011), 176.

34 Microwave News November 8, 2012.

Chapter 7: Electrohypersensitivity

1 *Possible Health Implications of Subjective Symptoms and Electromagnetic Fields*, a report prepared by a European group of experts for the European Commission, DGV. Arbete och Hälsa, 1997:19. Swedish National Institute for Working Life, Stockholm, Sweden.

2 Orjan Hallberg and Gerd Oberfield, 'Letter to the Editor: Will We All Become Electrosensitive?', *Electromagnetic Biology and Medicine*, 25 (2006), 189–191.

3 Dominique Belpomme, Christine Campagnac, and Philippe Irigaray, 'Reliable Disease Biomarkers Characterizing and Identifying Electrohypersensitivity and Multiple Chemical Sensitivity as Two Etiopathogenic Aspects of a Unique Pathological Disorder)', *Reviews on Environmental Health*, 30/4 (2015), 251–271.

4 Olle Johansson, 'Electrohypersensitivity: A Functional Impairment Due to an Inaccessible Environment', *Reviews on Environmental Health*, 30/4 (2015), 311–321.

Chapter 8: Hormone of Darkness

1 D. Acuña-Castroviejo, G. Escames, C. Venegas, M. E. Díaz-Casado, E. Lima-Cabello, L. C. López, S. Rosales-Corral, D. X. Tan, R. J. Reiter, 'Extrapineal Melatonin: Sources, Regulation, and Potential Functions', *Cellular and Molecular Life Sciences*, 71/16 (August 2014), 2997–3025.

2 A. J. Lewy, T. A. Wehr, F. K. Goodwin, D. A. Newsome, S. P. Markey, 'Light Suppresses Melatonin Secretion in Humans', *Science*, 210 (1980), 1267–1269.

3 J. J. Gooley, K. Chamberlain, K. A. Smith, S. B. Khalsa, S. M. Rajaratnam, E. Van Reen, J. M. Zeitzer, C. A. Czeisler, S. W. Lockley, 'Exposure to Room Light before Bedtime Suppresses Melatonin Onset and Shortens Melatonin Duration in Humans', *Journal of Clinical Endocrinology and Metabolism*, 96 (2011), E463–72.

4 C. Cajochen, S. Frey, D. Anders, J. Späti, M. Bues, A. Pross, et al., 'Evening Exposure to a Light-Emitting Diodes (LED)-Backlit Computer Screen Affects Circadian Physiology and Cognitive Performance', *Journal of Applied Physiology*, 110 (2011), 1432–8.

5 D. X. Tan, R. J. Reiter, L. C. Manchester, M. T. Yan, M. El-Sawi, R. M. Sainz, J. C. Mayo, R. Kohen, M. Allegra, R. Hardeland, 'Chemical and Physical Properties and Potential Mechanisms: Melatonin as a Broad Spectrum Antioxidant and Free Radical Scavenger,' *Current Topics in Medicinal Chemistry*, 2/2 (February 2002), 181–97.

6 B. Poeggeler, S. Thuermann, A. Dose, M. Schoenke, S. Burkhardt, R. Hardeland, 'Melatonin's Unique Radical Scavenging Properties—Roles of Its Functional Substituents as Revealed by a Comparison with Its Structural Analogs', *Journal of Pineal Research*, 33/1 (August 2002), 20–30.

7 D. X. Tan, L. C. Manchester, M. P. Terron, R. J. Reiter, 'One Molecule, Many Derivatives: A Never-Ending Interaction of Melatonin with Reactive Oxygen and Nitrogen Species?', *Journal of Pineal Research*, 42/1 (January 2007), 28–42.

8 M. N. Halgamuge, 'Critical Time Delay of the Pineal Melatonin Rhythm in Humans Due to Weak Electromagnetic Exposure', *Indian Journal of Biochemistry & Biophysics*, 50/4 (August 2003), 259–65.

9 D. L. Henshaw, J. P. Ward, J. C. Matthews, 'Can Disturbances in the Atmospheric Electric Field Created by Powerline Corona Ions Disrupt Melatonin Production in the Pineal Gland?' *Journal of Pineal Research*, 45/4 (November 2008), 341–50.

10 D. L. Henshaw, R. J. Reiter, 'Do Magnetic Fields Cause Increased Risk of Childhood Leukemia via Melatonin Disruption?', *Bioelectromagnetics*, Suppl 7 (2005), S86–97.

[11] L. Zhao, X. Liu, C. Wang, K. Yan, X. Lin, S. Li, H. Bao, X. Liu, 'Magnetic Fields Exposure and Childhood Leukemia Risk: A Meta-Analysis Based on 11,699 Cases and 13,194 Controls', *Leukemia Research*, 38/3 (March 2014), 269–74.

[12] S. I. Rapoport, T. K. Breus, 'Melatonin as a Most Important Factor of Natural Electromagnetic Fields Impacting Patients with Hypertensive Disease and Ccoronary Heart Disease', *Klinicheskaia Meditsina* (Mosk), 89/3 (2011), 9–14.

[13] Germaine Escames et al., 'The Role of Mitochondria in Brain Aging and the Effects of Melatonin', *Current Neuropharmacology*, 8/3 (8 September 2010), 182–193.

[14] Abha Srivastava, Yogesh Saxena, 'Effect of Mobile Usage on Serum Melatonin among Medical Students', *Indian Journal of Physiology and Pharmacology*, 58/4 (2014), 395–399.

[15] J. B. Burch, J. S. Reif, M. G. Yost, T. J. Keefe, C. A. Pitrat, 'Reduced Excretion of a Melatonin Metabolite in Workers Exposed to 60 Hz Magnetic Fields', *American Journal of Epidemiology*, 150/1 (1 July 1999), 27–36.

[16] Mohamed El-Helaly, E. Abu-Hashem, 'Oxidative Stress, Melatonin Level, and Sleep Insufficiency among Electronic Equipment Repairers', *Indian Journal of Occupational and Environmental Medicine*, 14/3 (2010), 66–70.

[17] R. Kc, X. Li, R. M. Voigt, M. B. Ellman, K. C. Summa, M. H. Vitaterna, A. Keshavarizian, F. W. Turek, Q. J. Meng, G. S. Stein, A. J. van Wijnen, D. Chen, C. B. Forsyth, H. J. Im., 'Environmental Disruption of Circadian Rhythm Predisposes Mice to Osteoarthritis-Like Changes in Knee Joint', *Journal of Cell Physiology*, 230/9 (September 2015), 2174–83.

[18] M. Gavella, V. Lipovac: Antioxidative Effect of Melatonin on Human Spermatozoa', *Archives of Andrology*, 44 (200), 23–7.

[19] K. Straif, R. Baan, Y. Grosse, B. Secretan, F. El Ghissassi, V. Bouvard, A. Altieri, L. Benbrahim-Tallaa, V. Cogliano, 'Carcinogenicity of Shift-Work, Painting, and Fire-Fighting', *Lancet Oncology*, 8/12 (December 2007), 1065–6.

[20] J. A. Lie, H. Kjuus, S. Zienolddiny, et al., 'Night Work and Breast Cancer Risk among Norwegian Nurses: Assessment by Different Exposure Metrics', *American Journal of Epidemiology*, 173 (2011), 1272–9.

[21] E. E. Flynn-Evans, R. G. Stevens, H. Tabandeh, E. S. Schernhammer, S. W. Lockley, 'Total Visual Blindness Is Protective against Breast Cancer', *Cancer Causes Control*, 20/9 November 2009), 1753–6.

Chapter 9: Genes Are under Attack

[1] H. Lai, N. P. Singh, 'Acute Low-Intensity Microwave Exposure Increases DNA Single-Strand Breaks in Rat Brain Cells', *Bioelectromagnetics*, 16/3 (1995), 207–10.

[2] H. Lai, N. P. Singh, 'Single- and Double-Strand DNA Breaks in Rat Brain Cells after Acute Exposure to Radiofrequency Electromagnetic Radiation', *International Journal of Radiation Biology*, 69/4 (April 1996), 513–21.

[3] H. Lai, N. P. Singh, 'Acute Exposure to a 60 Hz Magnetic Field Increases DNA Strand Breaks in Rat Brain Cells', *Bioelectromagnetics*, 18/2 (1997), 156–65.

[4] H. Lai, N. P. Singh, 'Melatonin and N-Tert-Butyl-Alpha-Phenylnitrone Block 60-Hz Magnetic Field-Induced DNA Single and Double Strand Breaks in Rat Brain Cells', *Journal of Pineal Research*, 22/3 (April 1997), 152–62.

[5] H. Lai, N. P. Singh, 'Melatonin and a Spin-Trap Compound Block Radiofrequency Electromagnetic Radiation-Induced DNA Strand Breaks in Rat Brain Cells', *Bioelectromagnetics*, 18/6 (1997), 446–5.

[6] H. Lai, N. P. Singh, 'Magnetic Field Induced DNA Strand Breaks in Brain Cells of the Rat', *Environmental Health Perspectives*, 112/6 (2004), 687–694.

[7] J. L. Phillips, N. P. Singh, and H. Lai, 'Electromagnetic Fields and DNA Damage', *Pathophysiology*, 16 (2009), 79–88.

[8] T. D. Whitehead, E. G. Moros, B. H. Brownstein, J. L. Roti Roti, 'Gene Expression Does Not Change Significantly in C3H 10T(1/2) Cells after Exposure to 847.74 CDMA or 835.62 FDMA Radiofrequency Radiation', *Radiat Research*, 165/6 (June 2006), 626–35.

[9] Vijayalaxmi, T. J. Prihoda, 'Genetic Damage in Human Cells Exposed to Non-Ionizing Radiofrequency Fields: A Meta-Analysis of the Data

from 88 Publications (1990–2011)', *Mutation Research*, 749/1–2 (12 December 2012), 1–16.

[10] R. S. Malyapa, E. W. Ahern, W. L. Straube, E. G. Moros, W. F. Pickard, J. L. Roti Roti, 'Measurement of DNA Damage after Exposure to 2450 MHz Electromagnetic Radiation', *Radiation Research*, 1481 (1997), 608–617.

[11] E. Diem, C. Schwarz, F. Adlkofer, O. Jahn, H. Rudiger, 'Non-Thermal DNA Breakage by Mobile-Phone Radiation (1800–MHz) in Human Fibroblasts and in Transformed GFSH-R17 Rat Granulosa Cells In Vitro', *Mutation Research*, 583 (2005), 178–183.

[12] G. Gandhi, Anita, 'Genetic Damage in Mobile Phone Users: Some Preliminary Findings', *Indian Journal of Human Genetics*, 11 (2005), 99–104.

[13] T. Nikolova, J. Czyz, A. Rolletschek, P. Blyszczuk, J. Fuchs, G. Jovtchev, J. Schuderer, N. Kuster, A. M. Wobus, 'Electromagnetic Fields Affect Transcript Levels of Apoptosis-Related Genes in Embryonic Stem Cell–Derived Neural Progenitor cells', *FASEB Journal*, 19 (2005), 1686–1688.

[14] S. Lixia, K. Yao, W. Kaijun, L. Deqiang, H. Huajun, G. Xiangwei, W. Baohong, Z. Wei, L. Jianling, W. Wei, 'Effects of 1.8-GHz Radiofrequency Field on DNA Damage and Expression of Heat Shock Protein 70 in Human Lens Epithelial Cells', *Mutation Research*, 602 (2006), 135–142.

[15] R. J. Aitken, L. E. Bennetts, D. Sawyer, A. M. Wiklendt, B. V. King, 'Impact of Radio Frequency Electromagnetic Radiation on DNA Integrity in the Male Germline', *International Journal of Andrology*, 28 (2005), 171–179.

[16] C. Calderón, D. Addison, T. Mee, R. Findlay, M. Maslanyj, E. Conil, H. Kromhout, A. K. Lee, M. R. Sim, M. Taki, N. Varsier, J. Wiart, E. Cardis, 'Assessment of Extremely Low Frequency Magnetic Field Exposure from GSM Mobile Phones', *Bioelectromagnetics*, 35/3 (April 2014), 210–21.

[17] T. Linde, K. H. Mild, 'Measurement of Low Frequency Magnetic Fields from Digital Cellular Telephones', *Bioelectromagnetics*, 18/2 (1997), 184–6.

[18] A. Hekmat, A. A. Saboury, A. A. Moosavi-Movahedi, 'The Toxic Effects of Mobile Phone Radiofrequency (940 MHz) on the Structure of Calf Thymus DNA', *Ecotoxicology and Environmental Safety*, 88 (February 2016), 35–41.

[19] E. Diem, C. Schwarz, F. Adlkofer, O. Jahn, H. Rüdiger, 'Non-Thermal DNA Breakage by Mobile-Phone Radiation (1800 MHz) in Human Fibroblasts and in Transformed GFSH-R17 Rat Granulosa Cells In Vitro', *Mutation Research*, 583/2 (6 June 2005), 178–83.

[20] R. Winker, S. Ivancsits, A. Pilger, F. Adlkofer, H. W. Rüdiger, 'Chromosomal Damage in Human Diploid Fibroblasts by Intermittent Exposure to Extremely Low-Frequency Electromagnetic Fields', *Mutation Research*, 585/1–2 (1 August 2005), 43–9.

[21] C. Schwarz, E. Kratochvil, A. Pilger, N. Kuster, F. Adlkofer, H. W. Rüdiger, 'Radiofrequency Electromagnetic Fields (UMTS, 1,950 MHz) Induce Genotoxic Effects In Vitro in Human Fibroblasts but Not in Lymphocytes', *International Archives of Occupational and Environtal Health*, 81/6 (May 2008), 755–67, doi: 10.1007/s00420-008-0305-5.

[22] S. Franzellitti, P. Valbonesi, N. Ciancaglini, C. Biondi, A. Contin, F. Bersani, E. Fabbri, 'Transient DNA Damage Induced by High-Frequency Electromagnetic Fields (GSM 1.8 GHz) in the Human Trophoblast HTR-8/SVneo Cell Line Evaluated with the Alkaline Comet Assay', *Mutation Research*, 683/1–2 (5 January 2010), 35–42.

[23] G. Guler, A. Tomruk, E. Ozgur, N. Seyhan, 'The Effect of Radiofrequency Radiation on DNA and Lipid Damage in Non-Pregnant and Pregnant Rabbits and Their Newborns', *General Physiology and Biophysics*, 29/1 (March 2010), 59–66.

[24] K. K. Kesari, J. Behari, S. Kumar, 'Mutagenic Response of 2.45 GHz Radiation Exposure on Rat Brain', *International Journal of Radiation Biology*, 86/4 (April 2010), 334–43.

[25] E. Karaca, B. Durmaz, H. Aktug, T. Yildiz, C. Guducu, M. Irgi, M. G. Koksal, F. Ozkinay, C. Gunduz, O. Cogulu, 'The Genotoxic Effect of Radiofrequency Waves on Mouse Brain', *Journal of Neuro-Oncology*, 106/1 (January 2012), 53–8.

[26] F. Mancinelli, M. Caraglia, A. Abbruzzese, G. d'Ambrosio, R. Massa, E. Bismuto, 'Non-Thermal Effects of Electromagnetic Fields at Mobile Phone Frequency on the Refolding of an Intracellular Protein:

Myoglobin', *Journal of Cellular Biochemistry*, 93/1 (1 September 2004), 188–96.

[27] Chuan Liu et al., 'Exposure to 1800 MHz Radiofrequency Electromagnetic Radiation Induces Oxidative DNA Base Damage in a Mouse Spermatocyte-Derived Cell Line', *Toxicology Letters*, 218/1 (2003), 2–9.

[28] L. Zotti-Martelli, M. Peccatori, V. Maggini, M. Ballardin, R. Barale, 'Individual Responsiveness to Induction of Micronuclei in Human Lymphocytes after Exposure In Vitro to 1800-MHz Microwave Radiation', *Mutation Research*, 582/1–2 (4 April 2005), 42–52.

[29] Eva Markovà, Lars O. G. Malmgren, and Igor Y. Belyaev, 'Microwaves from Mobile Phones Inhibit 53BP1 Focus Formation in Human Stem Cells More Strongly Than in Differentiated Cells: Possible Mechanistic Link to Cancer Risk', *Environmental Health Perspectives*, 118/3 (March 2010), 394–399.

[30] I. Belyaev, E. Markova, L. Malmgren, 'Microwaves from Mobile Phones Inhibit 53BP1 Focus Formation in Human Stem Cells Stronger Than in Differentiated Cells: Possible Mechanistic Link to Cancer Risk', *Environmental Health Perspectives*, 22 Oct 2009.

[31] S. Panier, S. J. Boulton, 'Double-Strand Break Repair: 53BP1 Comes into Focus', *Nature Reviews Molecular Cell Biology*, 15(1 January 2014), 7–18.

Chapter 10: Radiofrequency Causes Cellular Stress

[1] Denham Harman, 'Aging: A Theory Based on Free Radical and Radiation Chemistry', *Journal of Gerontology*, 11/3 (1956), 298–300.

[2] M. Simko et al., 'Extremely Low Frequency Electromagnetic Fields as Effectors of Cellular Responses In Vitro: Possible Immune Cell Activation', *Journal of Cellular Biochemistry*, 93 (2004), 83–92.

[3] W. A. Catterall, 'Structure and Regulation of Voltage-Gated Ca^{2+} Channels', *Annual Review of Cell and Developmental Biology*, 16 (2000), 521–55.

4 J. Walleczek, 'Electromagnetic Field Effects on Cells of the Immune System: The Role of Calcium Signaling', *FASEB Journal*, 6 (1992), 3177–85.

5 I. Yakymenko, O. Tsybulin, E. Sidorik, D. Henshel, O. Kyrylenko, S. Kyrylenko, 'Oxidative Mechanisms of Biological Activity of Low-Intensity Radiofrequency Radiation', *Electromagnetic Biology and Medicine*, (7 July 2015), 1–16.

6 M. L. Pall, 'Electromagnetic Fields Act via Activation of Voltage-Gated Calcium Channels to Produce Beneficial or Adverse Effects', *Journal of Cellular and Molecular Medicine*, 17/8 (August 2013), 958–65.

7 G. L. Craviso, S. Choe, P. Chatterjee, et al., 'Nanosecond Electric Pulses: A Novel Stimulus for Triggering Ca^{2+} Influx into Chromaffin Cells via Voltage-Gated Ca^{2+} Channels', *Cellular and Molecular Neurobiology*, 30 (2010), 1259–65.

8 A. A. Pilla, 'Electromagnetic Fields Instantaneously Modulate Nitric Oxide Signaling in Challenged Biological Systems', *Biochemical and Biophysical Research Communications*, 426/3 (28 September 2012), 330–3.

9 S. H. Francis, J. L. Busch, J. D. Corbin, et al., 'cGMP-Dependent Protein Kinases and cGMP Phosphodiesterases in Nitric Oxide and cGMP Action', *Pharmacological Reviews*, 62 (2010), 525–63.

10 S. V. Lymar, R. F. Khairutdinov, J. K. Hurst, 'Hydroxyl Radical Formation by O-O Bond Homolysis in Peroxynitrous Acid', *Inorganic Chemistry*, 42 (2003), 5259–66.

11 J. Gmitrov, C. Ohkuba, 'Verapamil Protective Effect on Natural and Artificial Magnetic Field Cardiovascular Impact', *Bioelectromagnetics*, 23 (2002), 531–41.

12 G. Szabó, S. Bährle, 'Role of Nitrosative Stress and Poly(ADP-Ribose) Polymerase Activation in Myocardial Reperfusion Injury', *Current Vascular Pharmacology*, 3 (2005), 215–20.

13 Y. Sakihama, M. Maeda, M. Hashimoto, et al., 'Beetroot Betalain Inhibits Peroxynitrite-Mediated Tyrosine Nitration and DNA Strand Damage', *Free Radical Research*, 46 (2012), 93–9.

14 A. C. Mannerling, M. Simko, K. H. Mild, M. O. Mattsson, 'Effects of 50-Hz Magnetic Field Exposure on Superoxide Radical Anion

Formation and HSP70 Induction in Human K562 Cells', *Radiation and Environmental Biophysics*, 49/4 (November 2012), 731–41.

15 R.J. Rozek, M.L. Sherman, A.R. Liboff, B.R. McLeod, S.D. Smith, 'Nifedipine in an Antagonist to Cyclotron Resonance Enhancement of Ca^{2+} incorporation in human Lymphocytes', *Cell Calcium*, 8/6 (1987), 413-427.

16 D. Lyle et al., 'Calcium Uptake by Leukemic and Normal T Lymphocytes Exposed to Low Frequency Magnetic Fields', *Biomagnetics*, 12 (1991), 145–156.

17 H. Lai, N. P. Singh, 'Single- and Double-Strand DNA Breaks in Rat Brain Cells after Acute Exposure to Radiofrequency Electromagnetic Radiation', *International Journal of Radiation Biology*, 69 (4 (April 1996), 513–21.

18 H. Lai, N. P. Singh, 'Magnetic-Field-Induced DNA Strand Breaks in Brain Cells of the Rat', *Environmental Health Perspectives*, 112/6 (May 2004), 687–94.

19 R. G. Stevens, 'Electromagnetic Fields and Free Radicals', *Environmental Health Persepctives*, 112/13 (September 2004).

20 M. Blank, R. Goodman, 'Electromagnetic Fields May Act Directly on DNA', *Journal of Cellular Biochemistry*, 75 (1997), 369–74.

21 K. K. Cheng et al., 'Exposure to Power Frequency Magnetic Fields and the Risk of Childhood Cancer', *Lancet*, 354 (1999), 1925–931.

22 H. Lai, N. P. Singh, 'Magnetic Field Induced DNA Strand Breaks in Brain Cells of the Rat', *Environmental Health Perspectives*, 112 (2004), 687–94.

23 Y. Luo et al., 'Oxidative Damage to DNA Constituents by Iron Mediated Fenton Reaction', *Journal of Biological Chemistry*, 271 (1996), 21167–76.

24 H. Lai, N. P. Singh, 'Melatonin and a Spin-Trap Compound Block RF EMG Radiation-Induced DNA Strand Breaks in Rat Brain Cells', *Bioelectromagnetics*, 18 (1997), 446–54.

25 H. Lai, N. P. Singh, 'Melatonin and N-Tert-Butyl-Alpha-Phenylnitrone Block 60-Hz Magnetic Field Induced Single and Double Stand Breaks in Rat Brain Cells', *Journal of Pineal Research*, 22 (1997), 152–62.

26 H. Lai, N. P. Singh, 'Single and Double Strand Breaks in Rat Brain Cells after Acute Exposure to RF EMG Radiation', *International Journal of Radiation Biology*, 69 (1996), 513–21.

[27] H. Lai, N. P. Singh, 'Acute Low Intensity MW Exposure Increases DNA Single-Strand Breaks in Rat Brain Cells', *Bioelectromagnetics*, 16 (1995), 207–210.

[28] J. Martínez-Sámano, P. V. Torres-Durán, M. A. Juárez-Oropeza, and L. Verdugo-Díaz, 'Effect of Acute Extremely Low Frequency Electromagnetic Field Exposure on the Antioxidant Status and Lipid Levels in Rat Brain,' *Archives of Medical Research*, 43/3 (2012), 183–189.

[29] M. Simkó, D. Richard, R. Kriehuber, D. G. Weiss, 'Micronucleus Induction in Syrian Hamster Embryo Cells Following Exposure to 50 Hz Magnetic Fields, Benzo(a)pyrene, and TPA In Vitro', *Mutattion Research*, 495/1–2 (22 August 2001), 43–50.

[30] A. Ilhan, A. Gurel, F. Armutcu, S. Kamisli, M. Iraz, O. Akyol, S. Ozen, 'Ginkgo Biloba Prevents Mobile Phone-Induced Oxidative Stress in Rat Brain', *Clinica Chimica Acta*, 340/1–2 (February 2004), 153–62.

[31] Antonia Patruno, Shams Tabrez, Mirko Pesce, Shazi Shakil, Mohammad A. Kamal, Marcella Reale, 'Effects of Extremely Low Frequency Electromagnetic Field (ELF – EMF) on Catalase, Cytochrome P450 and Nitric Oxide Synthase in Erythroleukaemic Cells', *Life Sciences*, 121 (15 January 2015), 117–123.

Chapter 11: Leaky Blood Brain Barrier

[1] S. R. Parathath, S. Parathath, S. E. Tsirka, 'Nitric Oxide Mediates Neurodegeneration and Breakdown of the BBB in tPA-Dependent Excitotoxic Injury in Mice', *Journal of Cell Science*, 119 (2006), 239–249.

[2] A. H. Frey, S. R. Feld, B. Frey, 'Neural Function and Behaviour: Defining the Relationship', *Annals of the New York Acaddemy of Sciences*, 247 (1975), 433–439.

[3] K. J. Oscar, T. D. Hawkins, 'Microwave Alteration of the BBB System of Rats', *Brain Research*, 126 (1977), 281–293.

[4] L. G. Salford, A. Brun, K. Sturesson, J. L. Eberhardt, B. R. Persson, 'Permeability of the Blood-Brain Barrier Induced by 915 MHz Electromagnetic Radiation, Continuous Wave and Modulated at 8, 16, 50, and 200 Hz', *Microscopy Research and Technique*, 27/6 (15 April 1994), 535–42.

[5] H. Nittby, A. Brun, J. Eberhardt, L. Malmgren, B. R. Persson, L. G. Salford, 'Increased Blood-Brain Barrier Permeability in Mammalian Brain 7 Days after Exposure to the Radiation from a GSM-900 Mobile Phone', *Pathophysiology*, 16/2–3 (August 2009), 103–12.

[6] L. G. Salford, A. Brun, J. L. Eberhardt, L. Malmgren, B. R. R. Persson, 'Nerve Cell Damage in Mammalian Brain after Exposure to Microwaves from GSM Mobile Phones, *Environmental Health Perspectives*, 111 (2003), 881–883.

[7] H. Nittby, G. Grafström, D. P. Tian, L. Malmgren, A. Brun, B. R. Persson, L. G. Salford, J. Eberhardt, 'Cognitive Impairment in Rats after Long-Term Exposure to GSM-900 Mobile Phone Radiation', *Bioelect@mrssoromagnetics*, 29/3 (April 2008), 219–32.

[8] B. R. R. Persson et al. (also with Salford from Sweden), 'Effects of Microwaves from GSM Mobile Phones on the Blood-Brain Barrier and Neurons in Rat Brain', *Progress in Electromagnetics Research Symposium 2005*, Hangzhou, China, August 22–26, 638–641.

[9] S. Balaguru, R. Uppal, R. P. Vaid, B. P. Kumar, 'Investigation of the Spinal Cord as a Natural Receptor Antenna for Incident Electromagnetic Waves and Possible Impact on the Central Nervous System', *Electromagnetic Biology and Medicine*, 31/2 (June 201), 101–11.

[10] J. Tang, Y. Zhang, L. Yang, Q. Chen, L. Tan, S. Zuo, H. Feng, Z. Chen, G. Zhu, 'Exposure to 900MHz Electromagnetic Fields Activates the mkp-1/ERK Pathway and causes Blood-Brain Barrier Damage and Cognitive Impairment in Rats', *Brain Research*, 1601 (March 2015), 92–101.

[11] Hao et al., 'Effects of Long-Term Electromagnetic Field Exposure on Spatial Learning and Memory in Rats', *Neurological Sciences*, 34/2 (2003), 157–64.

[12] O. Bas et al., '900 MHz Electromagnetic Field Exposure Affects Qualitative and Quantitative Features of Hippocampal Pyramidal Cells in the Adult Female Rat', *Brain Research*, 1265 (10 Apr 2009), 178–85.

[13] C. Chen et al., 'Exposure to 1800 MHz Radiofrequency Radiation Impairs Neurite Outgrowth of Embryonic Neural Stem Cells', *Scientific Reports*, 4 (29 May 2014), 5103.

[14] Q. Ma et al., 'Extremely Low-Frequency Electromagnetic Fields Affect Transcript Levels of Neuronal Differentiation-Related Genes in Embryonic Neural Stem Cells', *PLoS One*, 9/3 (3 March 2014).

[15] Y. C. Kuo, H. H. Chen, 'Effect of Electromagnetic Field on Endocytosis of Cationic Solid Lipid Nanoparticles by Human Brain-Microvascular Endothelial Cells', *Journal of Drug Targeting*, 18/6 (July 2010), 447–456.

[16] Y. C. Kuo, C. Y. Kuo, 'Electromagnetic Interference in the Permeability of Saquinavir across the Blood-Brain Barrier Using Nanoparticulate Carriers', *International Journal of Pharmaceutics*, 351/1–2 (3 March 2008), 271–81.

Chapter 12: Mobile Phones and Infertility

[1] R. J. Aitken and J. S. Clarkson, 'Cellular Basis of Defective Sperm Function and Its Association with the Genesis of Reactive Oxygen Species by Human Spermatozoa', *Journal of Reproduction and Fertility*, 81 (1987), 459–469.

[2] R. J. Aitken, T. B. Smith, M. S. Jobling, M. A. Baker, G. N. De Iuliis, 'Oxidative Stress and Male Reproductive Health', *Asian Journal of Andrology*, 16/1 (January–February 2014), 31–8.

[3] S. E. Lewis, R. J. Aitken, 'DNA Damage to Spermatozoa Has Impacts on Fertilization and Pregnancy', *Cell and Tissue Research*, 322 (2005), 33–41.

[4] A. Agarwal, R. A. Saleh, M. A. Bedaiwy, 'Role of Reactive Oxygen Species in the Pathophysiology of Human Reproduction', *Fertility and Sterility*, 79 (2003): 829–43.

[5] R. J. Aitken, M. A. Baker, 'Oxidative Stress, Sperm Survival and Fertility Control', *Molecular and Cellular Endocrinology*, 250 (2006), 66–9.

[6] A. J. Koppers, M. L. Garg, R. J. Aitken, 'Stimulation of Mitochondrial Reactive Oxygen Species Production by Unesterified, Unsaturated Fatty Acids in Defective Human Spermatozoa', *Free Radical Biology & Medicine*, 48/1 (1 January 2010), 112–9.

[7] A. J. Koppers, G. N. De Iuliis, J. M. Finnie, E. A. McLaughlin, R. J. Aitken, 'Significance of Mitochondrial Reactive Oxygen Species

in the Generation of Oxidative Stress in Spermatozoa', *Journal of Clinical Endocrinology and Metabolism*, 93/8 (2008), 3199–207.

8 K. Liu, Y. Li, G. Zhang, J. Liu, J. Cao, L. Ao, S. Zhang, 'Association between Mobile Phone Use and Semen Quality: A Systemic Review and Meta-Analysis', *Andrology*, 2/4 (2014), 491–501.

9 J. A. Adams, T. S. Galloway, D. Mondal, S. C. Esteves, F. Mathews, 'Effect of Mobile Telephones on Sperm Quality: A Systematic Review and Meta-Analysis', *Environmental International*, 70 (September 2014), 106–12.

10 S. La Vignera, R. A. Condorelli, E. Vicari, R. D'Agata, A. E. Calogero, 'Effects of the Exposure to Mobile Phones on Male Reproduction: A Review of the Literature', *Journal of Andrology*, 33/3 (May–June 2012), 350–6.

11 A. Agarwal, F. Deepinder, R. K. Sharma, G. Ranga, J. Li, 'Effect of Cell Phone Usage on Semen Analysis in Men Attending Infertility Clinic: An Observational Study', *Fertility and Sterility*, 89/1 (January 2008):124–8.

12 N. R. Desai, K. K. Kesari, A. Agarwal, 'Pathophysiology of Cell Phone Radiation: Oxidative Stress and Carcinogenesis with Focus on Male Reproductive System', *Reproductive Biology and Endocrinology*, 7 (2009), 114.

13 Adel Zalata, Ayman Z. El-Samanoudy, Dalia Shaalan, Youssef El-Baiomy, and Taymour Mostafa, 'In Vitro Effect of Cell Phone Radiation on Motility, DNA Fragmentation and Clusterin Gene Expression in Human Sperm', *International Journal of Fertility & Sterility*, 9/1 (April–June 2015), 129–136.

14 P. Strocchi, F. Rauzi, D. Cevolani, 'Neuronal Loss Up-Regulates Clusterin mRNA in Living Neurons and Glial Cells in the Rat Brain', *Neuroreport*, 10/8 (1999), 1789–1792.

15 S. Kumar, J. P. Nirala, J. Behari, R. Paulraj, 'Effect of Electromagnetic Irradiation Produced by 3G Mobile Phone on Male Rat Reproductive System in a Simulated Scenario', *Indian Journal of Experimental Biology*, 52/9 (September 2014), 890–7.

16 S. Roychoudhury, J. Jedlicka, V. Parkanyi, J. Rafay, L. Ondruska, P. Massanyi, J. Bulla, 'Influence of a 50 Hz Extra Low Frequency Electromagnetic Field on Spermatozoa Motility and Fertilization

Rates in Rabbits', *Journal of Environmental Science and Health, Part A. Toxic/Hazardous Substances & Environmental Engineering*, 44/10 (August 2009), 1041–7.

[17] S. K. Lee, S. Park, Y. M. Gimm, Y. W. Kim, 'Extremely Low Frequency Magnetic Fields Induce Spermatogenic Germ Cell Apoptosis: Possible Mechanism', *BioMed Research International*, 2014 (2014), 567183.

[18] Igor Gorpinchenko, Oleg Nikitin, Oleg Banyra, Alexander Shulyak, 'The Influence of Direct Mobile Phone Radiation on Sperm Quality,' *Central European Journal of Urology*, January 67 (2014), 65–71.

[19] B. M. Al-Ali, J. Patzak, K. Fischereder, K. Pummer, R. Shamloul, Cell Phone Usage and Erectile Function', *Central European Journal of Urology*, 66 (2013), 75–77.

[20] W. G. Whittow, C. J. Panagamuwa, M. A. Edwards, L. Ma, 'Indicative SAR Levels Due to an Active Mobile Phone in a Front Trouser Pocket in proximity to Common Metallic Objects', 2008 Loughborough Antennas and Propagation Conference in Loughborough, UK, 17–18 March, 149–152.

[21] C. Avendano, A. Mata, C. A. Sanchez Sarmiento, G. F. Doncel, 'Use of Laptop Computers Connected to Internet through Wi-Fi Decreases Human Sperm Motility and Increases Sperm DNA Fragmentation', *Fertility and Sterility*, 97 (2012), 39–45.

[22] S. S. Suarez, A. A. Pacey, 'Sperm Transport in the Female Reproductive Tract', *Human Reproduction Update*, 12 (2006), 23–37.

[23] Y. Yoshida, T. Seto, W. Ohsu, S. Hayashi, T. Okazawa, H. Nagase, M. Yoshida, H. Nakamura, '[Endocrine Mechanism of Placental Circulatory Disturbances Induced by Microwave in Pregnant Rats]', *Nihon Sanka Fujinka Gakkai Zasshi*, 47/2 (1995), 101–108.

Chapter 13: Heat Shock Proteins

[1] F. Ritossa, 'A New Puffing Pattern Induced by Temperature Shock and DNP in Drosophilia', *Experientia* 18 (1962), 571–573.

[2] Jong Youl Kim and Midori A. Yenari, 'The Immune Modulating Properties of the Heat Shock Proteins after Brain Injury', *Anatomy & Cell Biology*, 46/1 (March 2013), 1–7.

3 HSP 70 is involved in preventing neuronal apoptosis and opening up of BBB. S. Kelly, M. A. Yenari, 'Neuroprotection: Heat Shock Proteins', *Current Medical Research and Opinion*, 18 Suppl 2 (2002), s55–60.

4 H. Lin, M. Blank, R. Goodman, 'A Magnetic Field-Responsive Domain in the Human HSP70 Promoter', *Journal of Cellular Biochemistry*, 75/1 (1 October 1999), 170–6.

5 R. Goodman, M. Blank, 'Insights into Electromagnetic Interaction Mechanisms', *Journal of Cellular Physiology*, 192/1 (July 2002), 16–22.

6 A. C. Mannerling, M. Simkó, K. H. Mild, M. O. Mattsson, 'Effects of 50-Hz Magnetic Field Exposure on Superoxide Radical Anion Formation and HSP70 Induction in Human K562 Cells', *Radiation and Environmental Biophysics*, 49/4 (November 2010), 731–41.

7 K. Balakrishnan,V. Murali, C. Rathika, T. Manikandan, R. P. Malini, R. A. Kumar, M. Krishnan, 'Hsp70 Is an Independent Stress Marker among Frequent Users of Mobile Phones', *Journal of Environmental Pathology, Toxicology and Oncology*, 33/4 (2014), 339–347.

8 X. S. Yang, G. L. He, Y. T. Hao, Y. Xiao, C. H. Chen, G. B., Z. P. Yu, 'Exposure to 2.45 GHz Electromagnetic Fields Elicits an HSP-Related Stress Response in Rat Hippocampus', *Brain Research Bulletin*, 88/4 (2012), 371–378.

Chapter 14: Radiofrequency and Wildlife

1 R. Wiltschko, W. Wiltschko. 'Sensing Magnetic Directions in Birds: Radical Pair Processes Involving Cryptochrome', *Biosensors* (Basel), 4/3 (24 July 2014), 221–42.

2 K. Stapput, P. Thalau, R. Wiltschko, W. Wiltschko, 'Orientation of Birds in Total Darkness', *Current Biology*, 18/8 (22 April 2008), 602–6.

3 T. Ritz, P. Thalau, J. B. Phillips, R. Wiltschko, W. Wiltschko, 'Resonance Effects Indicate a Radical-Pair Mechanism for Avian Magnetic Compass', *Nature*, 429/6988 (13 May 2004), 177–80.

4 S. Engels, N. L. Schneider, N. Lefeldt, C. M. Hein, M. Zapka, A. Michalik, D. Elbers, A. Kittel, P. J. Hore, H. Mouritsen, 'Anthropogenic Electromagnetic Noise Disrupts Magnetic Compass Orientation in a Migratory Bird', *Nature*, 509/7500 (15 May 2014), 353–6.

[5] S. Cucurachi et al., 'A Review of the Ecological Effects of Radiofrequency Electromagnetic Fields (RF-EMF)', *Environment International*, 51 (2013), 116–140.

[6] D. Vanengelsdorp, M. D. Meixner, 'A Historical Review of Managed Honey Bee Populations in Europe and the United States and the Factors That May Affect Them', *Journal of Invertebrate Pathology*, 103 Suppl 1 (January 2010), S80–95.

[7] J. Hamzelou, 'Where Have All the Bees Gone?', *Lancet*, 370/9588 (25 August 2007), 639.

[8] G. Wellenstein, 'The Influence of High Tension Lines on Honeybee Colonies', *Z. Angew Entomology*, 74 (1973), 86–94.

[9] Ved Prakash Sharma and Neelima R. Kumar, 'Changes in Honeybee Behaviour and Biology under the Influence of Cellphone Radiations', *Current Science*, 98/10 (25 May 2010), 1376–78.

[10] Hart et al., 'Dogs Are Sensitive to Small Variations of the Earth's Magnetic Field', *Frontiers in Zoology*, 10 (2013), 80.

[11] A. Balmori, O. Hallberg, 'The Urban Decline of the House Sparrow (*Passer domesticus*): A possible Link with Electromagnetic Radiation', *Electromagnetic Biology and Medicine*, 26/2 (2007), 141–51.

[12] J. Everaert, D. Bauwens, 'A Possible Effect of Electromagnetic Radiation from Mobile Phone Base Stations on the Number of Breeding House Sparrows (*Passer domesticus*)', *Electromagnetic Biology and Medicine*, 26 (2007), 63–72.

[13] A. DiCarlo, N. White, F. Guo, P. Garrett, and T. Litovitz, 'Chronic Electromagnetic Field Exposure Decreases HSP70 Levels and Lowers Cytoprotection', *Journal of Cellular Biochemistry*, 84 (2002), 447–454.

[14] I. Grigor'ev, 'Biological Effects of Mobile Phone Electromagnetic Field on Chick Embryo (Risk Assessment Using the Mortality Rate)', *Radiatsionnaia Biologiia Radioecologiia*, 43/5 (2003), 541–3.

[15] T. D. Xenos and I. N. Magras. 'Low Power density RF radiation effects on experimental animal embryos and foetuses,' in P. Stavroulakis (ed.), *Biological Effects of Electromagnetic Fields* (Springer, 2003), 579–602.

[16] K. J. Lohmann, P. Luschi, G. C. Hays, 'Goal Navigation and Island-Finding in Sea Turtles', *Journal of Experimental Marine Biology and Ecology*, 356 (2008), 83–95.

[17] A. Balmori, 'Electromagnetic Pollution from Phone Masts. Effects on Wildlife', *Pathophysiology* (2009).

[18] David Roux et al., 'High Frequency (900 MHz) Low Amplitude (5 V m–1) Electromagnetic Field: A Genuine Environmental Stimulus That Affects Transcription, Translation, Calcium, and Energy Charge in Tomato', *Planta*, 227/4 (March 2008), 883–891.

[19] Rim A. Hussein, Magda A. El-Maghraby, 'Effect of Two Brands of Cell Phone on Germination Rate and Seedling of Wheat (*Triticum aestivum*)', *Journal of Environment Pollution and Human Health*, 2/4 (2014), 85–90.

Index

C

D

E

I

J

K

L

M

N

O

P

R

S

T

Printed in the United States
By Bookmasters